Beyond Relativism

ROGER D. MASTERS

Beyond
Relativism

SCIENCE AND HUMAN VALUES

DARTMOUTH COLLEGE

Published by University Press of New England / Hanover & London

Dartmouth College

Published by University Press of New England, Hanover, NH 03755

© 1993 by Roger D. Masters
Printed in the United States of America

5   4   3   2   1

CIP data appear at the end of the book

Contents

.

Preface

Today scientific explanations of the world often seem unrelated to the concerns of the average citizen. In our schools and colleges, we ask students to take courses from the natural sciences, the social sciences, and the humanities, yet we rarely show them how these disparate approaches fit together. Our newsmagazines and daily papers report scientific discoveries but usually fail to consider how these findings should guide practical decisions.

As a consequence, science is often mysterious and threatening for the public at large. Critics of the modern university complain that there is no concern for "values" but do not indicate how to introduce them without destroying the scientific enterprise on which a technological civilization rests.

This book is an attempt to make sense of the competing perspectives in contemporary natural science, traditional philosophy, and the study of human behavior. It is intended to make accessible, to a broad audience, the experience of an interdisciplinary course at an outstanding college or university. I hope to bring together the approaches and discoveries of what have become three divisions (natural science, social science, and humanities) that quite literally divide our institutions of higher learning.

The natural and social sciences seem concerned with facts: how do we explain the world around us? The humanities focus on the study of great thinkers and writers to gain perspective on moral and political choices: how should we live? For too long, these questions have been separated. The resulting belief that there is no objective or scientific foundation for moral and political values engenders relativism.

I challenge this position: the way we ought to live depends on the way we explain the world around us. It is necessary to go beyond the relativism that has dominated academic life during the twentieth century. Many people have sensed this and have been discouraged if not angered by the gap between science and morality. The resulting hostility to universities and intellectuals is paradoxical (and dangerous) in a society whose economic health and military power are so dependent on technology and science.

My book is an essay. It does not present a system of thought that can be digested easily. Rather, it is—like a good course at a fine liberal arts college like Dartmouth—a way of introducing issues and information that the intelligent citizen needs to know. As my teacher Leo Strauss said, such a course needs to be given on the assumption that "there is a silent student in class who knows more than you do." This work is addressed to anyone, from the college graduate (or undergraduate) to the scholar, who wonders how contemporary science can be related to human problems.

Too often, popularizations of science talk down to readers. This is both unnecessary and unwise. There is now a large public of intelligent and informed citizens. A higher proportion of young men and women pursue a college education than ever before, yet the media often fail to take these readers seriously. More important, the leaders and citizens of contemporary society need to confront the difficult issues continually generated by the latest science and technology.

My book therefore must address two audiences with apparently contradictory expectations: a general readership of informed adults or students seeking to make sense out of diverse scholarly disciplines, and the scholars who themselves work in these competing fields of study. To meet the different needs of these two groups, the main argument is presented in the text, with references and additional technical evidence placed in notes at the end of the volume.

My attempt to link contemporary science with the traditional understanding of human nature and morality is highly personal yet could not have been done alone. Having taught and written with scholars from other disciplines, I am immensely grateful to those who have shared their knowledge and insights. But neither these individuals nor the institutions that have supported my research (see Acknowledgments) bear responsibility for my conclusions.

This essay continues a tradition of intellectual inquiry that can be traced to Plato and Aristotle. Whether it succeeds depends greatly on the judgment of readers, much as the success of a classroom teacher depends greatly on the students. What we remember from our best teachers are the issues on which they focused and the skills they communicated to us. As in any good college course, the questions matter more than the answers. It is a good thing to remember that Socrates was considered the wisest man in ancient Greece because he knew that he didn't know.

South Woodstock, Vermont   R.D.M.
October 1992

Acknowledgments

It is with gratitude that I recognize the essential role of the institutions that have provided the material possibility for my scholarly research and writing and the individuals whose insight, friendship, and dialogue have taught me so much over the years. Without the support of the Simon Guggenheim Fellowship in 1967–68, the earlier grant from the Social Science Research Council and Yale University for my leave in 1963–64, and the continued funding of many organizations (the Harry Frank Guggenheim Foundation, the National Science Foundation, the National Endowment for the Humanities, the Maison des Sciences de l'Homme in Paris, the Rockefeller Center for the Social Sciences at Dartmouth, the Gruter Institute for Law and Behavioral Research), my career would have been impossible.

Many individuals have also been of immeasurable help to me: scholarship and scientific research—particularly if it involves experimentation—cannot be conducted by an isolated scholar. Of my teachers, none had as deep and lasting an impact as Leo Strauss, whose presence in the classroom had no equal and who may well have been the most important political thinker of his time. Of the many others at Harvard and the University of Chicago who shaped my understanding and increased my intellectual skills, I have special debts to Joseph Palamountain (who directed my undergraduate thesis and convinced me to follow an academic career) and Joseph Cropsey (whose insightful advice on my doctoral dissertation was the prelude to continuing friendship).

Of those who have shared my commitment to an understanding of the

Western philosophic tradition, I particularly thank the late Allan Bloom, Robert Faulkner, Father Ernest Fortin, Nanerl Keohane, Harvey Mansfield, Jr., Terrence Marshall, Heinrich Meier, James Bernard Murphy, Michael Platt, Michael Rosano, Arlene Saxonhouse, John Scott, and Vincent Starzinger. It was the late John Roddam who first directed my attention to the importance of Robert Ardrey's popularization of contemporary research on animal behavior and convinced me to study this scientific approach to what philosophers called the state of nature; after I'd begun to do so, Ardrey himself was exceptionally generous in recognizing the relevance of my questions concerning his understanding of ethology (thereby spurring me to further research of my own).

In my studies in political theory, the thought of Jean-Jacques Rousseau has had a prominent place, and in preparing the *Collected Writings of Rousseau*, I have learned immensely from my co-editor, Christopher Kelly. Throughout the years, I have had unmeasured pleasure from my abiding friendship with Henry Ehrmann, whose willingness to listen to my rambling thoughts has often helped me to clarify them.

The experimental research described in this book benefited enormously from my collaboration with Denis Sullivan (from whom I learned much when teaching together as well as in the decade of our joint research): his emphasis on intellectual rigor has been a continual reminder of the need to submit one's insights, no matter how intriguing, to the tests of fact and logic. I have also gained deeply from the judgment, knowledge, and critical acuity of many other colleagues at Dartmouth. In addition to my research collaboration with John Lanzetta and his colleagues in the Department of Psychology, I have had the pleasure of teaching with and learning from Joe Harris in Physics, Ken Korey in Anthropology, and Ed Berger and Thom Roos in Biology. And, it needs to be added, I have been exceptionally fortunate in the many excellent Dartmouth students who have taught me much whenever I have tried to teach them; in addition to those whose interest in political philosophy has led to thoughtful explorations of texts, a word of gratitude is necessary for those who have worked as assistants in my experimental studies. Among recent students, I have special debts to those whose Honors Theses contributed greatly to my work: Elise Plate, James Newton, John Scott, Stephen J. Carlotti, Jr., A. Michael Warnecke, and David Clancy.

Scholars and administrators attached to other institutions have also been of great assistance to me at various times. Thanks to Clemens Heller and Elina Almasy of the Maison des Sciences de l'Homme in Paris, in the years after 1971 I had the opportunity to meet Lionel Tiger, John Hurrell Crook, Mario Von Cranach, and others who have helped in editing the "Biology and Social Life" section of *Social Science Information*; in 1985, another research appointment at the Maison des Sciences de l'Homme led to a lasting collaboration with Siegfried Frey of the University of Duisburg, who, with his colleagues Gary Bente and Guido Kempter, have made it possible for me to go farther in cross-cultural research than I ever imagined was possible.

Since 1981, my work has also been immeasurably enriched by the Gruter Institute for Law and Behavioral Research, whose remarkable founder, Dr. Margaret Gruter, has played such a formidable role in expanding the understanding of the life sciences in the fields of law and the social sciences. Among those associated with this group, I am particularly indebted to Michael T. McGuire, E. Donald Elliott, Robin Fox, Robert Frank, Gordon Getty, William Rodgers Jr., and Lionel Tiger. At the same time, I have benefited especially from collaborating with Glendon Schubert as well as Albert Somit, the late Thomas Wiegele, and others in the Association for Politics and the Life Sciences.

Last but not least, two readers have helped to render my prose more intelligible: my wife, Susanne Masters (who has an unerring eye for confused prose), and my editor at the University Press of New England, Michael Lowenthal (who is equally adept at identifying unnecessary verbiage). To them, as to all of the others mentioned above or inadvertently omitted, my sincere thanks. None is responsible for what I have said, but all deserve my gratitude for helping me to say it.

R.D.M.

Yet Nature is made better by no mean
But Nature makes that mean; so over that art,
Which you say adds to Nature, is an art,
That Nature makes.

Shakespeare, *The Winter's Tale,*
IV.iv.89–92

PART ONE

THE NATURE OF SCIENCE

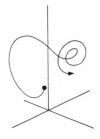

1. The Crisis of Modern Science and Society

As we approach the third millennium, technologically advanced society faces a crisis. Never before has so much been known about such an amazing array of natural things. Never before has a larger proportion of society completed higher education with a pretense of being able to comprehend these advances. And yet, in place of the optimism and belief in progress that has dominated Western opinion over the past 200 years, there is widespread pessimism, fear, and conflict. Why does the modern scientific project seem so fragile to those who apparently support and benefit from it?

This crisis reflects a growing tension between our unprecedented scientific knowledge and popular opinions about the world and human history. Both the general public and its leaders often lack an accurate understanding of science. Our schoolchildren are poorly trained in scientific subjects; average citizens know little about contemporary scientific theories and have difficulty evaluating scientific controversies; the urge for instant solutions deforms the description of new scientific findings and misdirects the focus of research; policy issues arising from science and technology are often reduced to ideological slogans; the funding and administration of science itself are frequently based on bureaucratic or political criteria that threaten to stifle creativity.[1]

Our mass media devote enormous attention to sports and popular culture, relatively little to medicine and technology, and almost none to pure science itself. Scientists are expected to be infallible and technology to be flawless. The slightest signs of scientific fraud, incompetence, or technical failure generate global skepticism about the entire process (reflecting child-ish impatience more than understanding of the actual process of scientific discovery and natural complexity). Our universities and other institutions of modern scientific activity have been criticized strongly for their failure to sustain the "values" on which Western civilizations rests.[2]

We benefit from the fruits of science and technology, but most of us look elsewhere for meaning in our lives: to material pleasure, to religion, to art and popular culture, to economic success, or to political activism. Science and the technical wonders it spawns are the basis of the success of our civilization, yet most citizens and not a few leaders find scientific matters forbiddingly difficult if not terrifying. Large sums of money are spent to send the next generation to colleges where modern science is taught; par-ents pay the bills without imagining that they themselves can and should understand the scientific basis of our civilization.

How could it be that the very success of modern science has led to such doubts, hostility, and ignorance? Why have we been able to create the greatest scientific community in human history while having such limited success in teaching the public (or even the elite) to appreciate what it does? To be sure, our society faces awesome problems: nuclear weapons, environ-mental pollution, economic insecurity, ethnic violence and war, poverty, disease, unprincipled greed—it is easy to list reasons for concern. But many of these reasons for fear are ancient, and all were present thirty years ago, when public opinions were far more optimistic than today. The roots of the crisis seem to be deeper, in the inability of modern science to address the ends or purposes to which its fruits are directed.

Since the seventeenth century, the modern scientific project has emerged as a very particular method of understanding nature: scientists are trained to discover how things work or why they happen but not what is good or bad. Philosophers call this value-free science. A gulf between the *Is* and the *Ought*—called the fact–value dichotomy—is said to prevent the sci-entist, who studies questions of fact, from addressing social or personal values as scientific questions: if a scientist raises ethical or political con-cerns, it is supposedly as a citizen, not as a specialist. Such a self-limiting restraint on science seemed to be the condition of replacing arid theological or speculative controversies with rigorous theories and empirical verifica-tion. Now, however, it is evident that this quest for scientific objectivity has had paradoxical consequences.[3]

The very success of a value-free science has produced intractable politi-cal and ethical problems. We *can* build hydrogen bombs and nuclear power stations. But *should* we? We *can* sequence the human genome and engage in genetic engineering. But *should* we? We *can* discover the mechanisms by which the body functions and use this knowledge to control reproduction or postpone death by technologies ranging from contraception and abortion to artificial organs, transplants, and respirators. But *should* we use all of these techniques, some of them, or none?

Two kinds of issues arise in this context, one practical and the other theoretical. In practice, we ask questions of public policy, ethics, and economics—and wonder how to direct, control, and finance the application of scientific or technical discoveries. A laboratory in France develops and sells a medication called RU486 that terminates pregnancy: why shouldn't it be legalized (or even tested) in the United States? A comatose woman has survived for seven years solely due to artificial means: should these life-support systems be disconnected? Technical measurements show that the earth's ozone layer is being depleted and trace the effect to specific chemical compounds: who should pay the added costs of alternatives? The political, ethical, and economic issues are endless, passionately controversial, and often confusing.

Perhaps we can gain deeper insight by focusing on the underlying problem from which all such issues arise. In theory, the crisis of modern science can be traced to the divorce between fact and value. A value-free science concerns means, not ends. How, then, are the ends to be chosen? Where will standards of right and wrong come from: religion? historical progress? social convention or custom? political power and coercion? arbitrary whim or personal choice? Can science say nothing whatever about the difference between good and bad, justice and injustice, or even reasonable and unreasonable uses of science itself?

For many contemporary philosophers and scientists, the attempt to derive preferences or moral standards from the natural sciences is called the "naturalistic fallacy."[4] Where does this belief come from? Is there good evidence to support the fact–value dichotomy on which it rests? To answer these questions, it is necessary to reconsider the widely shared assumption that science is—or should be—value-free. This is difficult because it requires an integration of scholarly disciplines and perspectives that usually have been isolated from each other.

There are two main reasons to combine scientific and humanistic perspectives. On the one hand, the conception of a value-free science needs to be reconsidered in the light of modern scientific discoveries. Presumably, our understanding of scientific method is not contradicted by science's results. As a result, broad theoretical questions need to be addressed on the basis of scientific evidence. On the other hand, the presumed gulf between fact and value needs to be related to the Western philosophic tradition and especially to the ancient Greek thinkers, epitomized by Plato and Aristotle, who encountered and rejected related views in their own day. The prevailing opinions of today's society therefore need to be reconsidered in the light of *both* contemporary scientific research *and* ancient philosophy.

It is to this ambitious but pressing task, which could be described as a study of the limitations of science,[5] that this book is directed. Elsewhere, I have tried to show that perennial issues of political and moral philosophy are illuminated by contemporary evolutionary biology.[6] Many other scholars have also argued that a "naturalist" approach to moral, ethical, and political principles provides a needed alternative to the relativism and nihilism of contemporary or postmodern intellectual life.[7] Such endeavors have raised the question of how far scientific research can contribute to a definition of standards of right and wrong.

This issue is important for the survival of our political institutions, of our species, and indeed of life itself. In the age of nuclear or biochemical war, genetic engineering, and global warming, the optimistic belief in scientific or technological solutions to all human problems is increasingly challenged. At no time, therefore, has it been more important to gain a deeper understanding of the implicit assumptions and techniques of modern science and to ask whether this way of understanding the world might not have some inherent limits or defects that have been generally ignored over the past three and a half centuries.[8]

My argument begins with a description of what can be called "modern science"—the project, first fully articulated by Sir Francis Bacon and elaborated in the sixteenth and seventeenth centuries, of using human knowledge to achieve the "conquest of nature" for the "relief of man's estate." In so doing, I will contrast this way of knowing with both religious faith (epitomized by the Judeo-Christian belief in one God) and the science of the ancient Greeks and Romans.

These three perspectives are distinct but related in complex ways. Each derives moral obligation from a different source: God's will, nature, or custom and education. Both ancient and modern science rely on human reason; revealed religion relies primarily on faith. But neither revealed religion nor ancient science envisages the gulf between fact and value that characterizes modern science. Or, to put it differently, secular thought today has substituted considerations of utility or efficiency for the emphasis on virtue shared by ancient philosophers and Judeo-Christian doctrine.[9] Hence, these three ways of knowing about the world can be visualized as forming a triangle, none of whose poles can be entirely reduced to the others (Fig. 1.1).[10]

To illuminate our predicament, I then focus more precisely on the specifically modern view of ends, or values. Why do so many contemporary

FIGURE I.I
Three Ways of Knowing

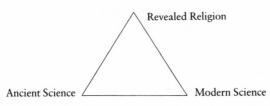

	Revealed Religion	Ancient Science	Modern Science
Source of knowledge	God	Nature	Education
Origin of human values	Faith	Reason	Reason
Standard of human life	Virtue (faith, hope, charity)	Virtue (wisdom, courage, justice, moderation)	Utility (efficiency, equity, progress)

scientists and intellectuals, unlike either pagan philosophers or Christian theologians, think that standards of judgment—the ends, purposes, or goals of human life—are so radically different from the "objective" phenomena of the physical and biological world?

While each of the three modes of knowledge depicted in Figure 1.1 has unique characteristics, there is an important difference between revealed religion and either ancient or modern science. Both ancient and modern science share a reliance on human reason, whereas the acceptance of divine revelation rests on religious belief. Neither philosophic exploration nor scientific evidence is likely to persuade fundamentalists to abandon their convictions; likewise, theological doctrine is not likely to change the opinions of the agnostic or atheistic scientist.

By contrasting ancient and modern science, we can see more clearly the "permanent" or inherent limitations on contemporary science while avoiding the intractable conflicts between faith and reason. To this end, I will examine the conventional arguments about the methods and limitations of modern science in the light of substantive scientific research in evolutionary biology, human cognition, and mathematics.

For example, although generations of philosophers and social scientists have accepted Locke's critique of "innate ideas," recent findings in neuroscience and nonverbal communication contradict the notion that the human brain is a tabula rasa, or blank slate. The age-old debates concerning the role of "nature" and "nuture" in human behavior are illuminated by new findings in developmental psychology, behavioral ecology, and human genetics. Modern science typically relies on evidence, but recent scientific evidence challenges widely held views of science itself. It is therefore time to reconsider alternative models of human knowledge.

Modern scientific method teaches us to abandon theories, paradigms, or assumptions when they are contradicted by the evidence, and I will argue that the evidence points to the validity of ancient science when properly understood. If the "naturalistic fallacy" is, from a scientific perspective, misleading as a way of understanding the relationship between means and ends, perhaps we can return to reasoned dialogue about the purposes or values guiding human life without being guilty of logical contradiction, wishful thinking, naïveté, ideology, or religious prejudice.

In focusing on the difference between the science of our own time and the understanding of thinkers from Thales and Socrates to Plato, Aristotle, and other ancients, I will stress epistemology (how a thing is known), mathematics (what time, form, and number are), and biology (what we know about animal and human nature), as well as the status of morality and duty (how obligation arises and what it entails). On this basis, I contend that the mood generally described as postmodernism—a relativistic or nihilistic conception that humans are alone in an alien universe with no inherent principles of value or meaning—is both intellectually incomplete and morally dangerous.

These considerations suggest the wisdom of the ancient view that nature, understood through reason, can provide a deeper understanding of human life and its proper ends or goals. This naturalist position differs from both dogmatic absolutes as well as from relativism and nihilism. Circumstances and history matter for living things. Human nature establishes ends and criteria for judgment but without providing simple, universal answers. In the tradition of ancient science, prudence is therefore necessary to apply theoretical knowledge to practical matters.

In stating that the modern scientific method has inherent limitations even for studying natural phenomena, I do not claim to resolve the conflict between reason and revelation; the issues separating human knowledge and divine grace are too deep—and, as contemporary events around the world testify, too explosive—for resolution in the present context. But in proposing that ancient science provides a naturalistic alternative to the relativism associated with modern science, it should not be forgotten that the scientific project that has dominated Western civilization since the Renaissance can also be challenged from the perspective of religious faith.

The organization of this essay is as follows. The remaining chapters in part 1 explore the difference between ancient and modern science. Chapter 2 outlines the history of the modern scientific project and its unique approach to knowledge and technology. In chapter 3, the difference between ancient and modern science is examined with regard to basic concepts of time, form, and knowledge.

In this discussion, a consideration of contemporary mathematical and physical theories reveals that, of the two main traditions in Western science, it is the ancient naturalism epitomized by Plato and Aristotle that is the more general. Modern science is powerful and accurate because it explores a limited domain of phenomena; the methods and theories we associate with a value-free science are not so much wrong as partial and incomplete. Thus one can understand the limitation of each approach to scientific knowledge without rejecting the possibility of science itself.

Part 2 turns to the problems of facts and values in more detail, addressing the claim of some contemporaries that rational or scientific knowledge about values is impossible. Chapter 4 considers the role of subjectivity in ancient and modern science, showing that dialogue is the only method by which intuition, reasoning, and experimental science can be integrated. But, it will be asked, how can humans communicate with each other without relying on the arbitrary conventions of language and culture?

Chapter 5 addresses this central problem. At the origins of modernity, naturalism was challenged by Locke and Hume on the grounds that innate ideas are nonexistent: people differ so radically in their responses to events that the methods of modern science seem necessary to establish even minimal areas of "intersubjective verifiability." Chapter 5 answers this criticism of ancient science by exploring the foundations of social communication, focusing on experimental findings about nonverbal displays and emotions as innate social signals.

The evolutionary origins of human behavior, the structure and function of the brain, and experimental studies of varied responses to leaders contradict the modern assumption that shared meaning is entirely due to custom and convention. Because social communication has a natural basis, it is possible to demonstrate the existence of innate ideas and to show how they are shaped by both culture and individual experience. Even among some nonhuman primates, moreover, these nonverbal cues and emotional responses are associated with "moralistic aggression"—the natural foundation of what, in humans, is called the sense of justice or injustice.[11]

Even if one grants that communication is possible, however, it does not follow that scientific reasoning is immune from the self-interest of the scientist. Chapter 6 therefore considers in detail the argument that all science is relative to the historical circumstances in which it develops. Such an examination is rendered particularly important by the claim that a biological or naturalist view of human society represents an ideological commitment to conservative or "right wing" political principles—a widespread belief that is not confirmed by careful study of theories in evolutionary biology.

In part 3, I will then show how the new naturalism (or "critical naturalism") makes it possible to relate facts and values. Chapter 7 treats the conception of value-free science in Locke, Hume, and Arnold Brecht, showing that the most frequently cited arguments against deriving values from facts are contradicted by recent scientific findings. Chapter 8 relates this negative argument to the positive one of elaborating the scientific foundations of ethical and political judgment by focusing on cognitive neuroscience, behavioral ecology, and mathematical theories of chaos. For each of these disciplines, I examine the theoretical concepts used to divorce values from facts, survey recent scientific evidence to show how it contradicts conventional relativism, and explore the implications for moral and political values. Chapter 9 then summarizes the understanding of human nature that follows from the research described throughout this book.

In the Conclusion, I assess this new naturalism, in which science and reason provide the foundation for standards of morality, justice, and sound public policy. By setting forth the logic of science according to the tradition of Plato and Aristotle, an approach to facts and values is outlined that avoids both the nihilism of postmodernism and the claims to absolute certainty of totalitarian ideologies and intolerant religious dogmas.

Theories of knowledge and politics that treat humans as isolated individuals without an evolutionary past are clearly inadequate. If so, the naturalism of the ancients seems to provide a viable alternative to the modern view that human life is entirely based on conventional or relative standards of truth and morality. Such a perspective should enable us to address the deeply felt malaise that threatens modern science and the society it has supported without falling into the extremes associated with authoritarian governments and arbitrary doctrines.

The rediscovery of a natural basis for meaning and value does not prove that moderns in the tradition of Leonardo, Machiavelli, Bacon, Locke, and Hume were totally incorrect or foolish to emphasize human creative power and convention: within limits, modern science works as effectively in politics as in the study of physics or chemistry. Nor can either ancient or modern science totally absorb or replace the claims of faith. Ultimately, the principles of convention, reason, and faith—representing the perspectives of modern science, ancient science, and revealed religion—must coexist.

The Western tradition has been unique because of its reliance on all three of these distinct approaches to knowledge and judgment. The attempt to integrate and balance the three apparently contradictory modes of thought is not impossible. The thought of Pascal, the great French mathematician who both conceived of something like the modern computer and expressed the deepest religious faith, suggests there may be a way out of the contemporary predicament. Perhaps it is possible to balance traditional perspectives in philosophy and religion with more modern scientific understanding. If so, we could find a humbler but sounder perspective than the hubris of a modern science cut off from its roots and claiming to "conquer nature" without principles on which to do so. Without such an approach, it is hard to see how the crisis of modern science can be overcome before either humanity or nature itself has been destroyed.

2. Modern Science: The Triumph of Method and the Conquest of Nature

Western civilization is unique in at least one respect. Our culture is the only one in human history that has dominated the entire planet. This integration of populations throughout the globe is due to our technology. And industrial technology is, in turn, derived from modern science. To understand contemporary society, we need to understand a kind of science that no other human civilization was able to develop.

What Makes Modern Science Different?

What makes the scientific approach to the world different from either religious faith or the science of antiquity? While philosophers of science often debate exactly what is meant by the "scientific method,"[1] its main features can readily be identified by contrasting all versions of contemporary science with the practices of ancient philosophers and medieval theologians.[2] That is, we can best see what makes modern science unique by emphasizing what is *not* characteristic of this form of knowing.

First, modern science does not derive knowledge from an authority who has a special claim to know the essence of things. In many traditions, the truth of a statement depends on the individual who utters it: the shaman, guru, or teacher "knows" because he or she has the authority to settle a dispute, to understand the omens, or to interpret the canonical texts. Modern scientists, in contrast, do not accept Einstein's answer in preference to that of a little-known physicist merely because of his prestige or office. To be sure, reputation can give a scholar an advantage: there is a politics of science as well as a rhetoric of science.[3] But while scientists often defer to status and convention, much as do ordinary people in local communities, authority is not explicitly *legitimate* as the basis of scientific knowledge. Whereas the laws of a state and the doctrines of a religion often depend on the authorities who approve them, scientists themselves feel guilty whenever it can be shown that their theories or empirical findings were accepted merely because of the author who presented them.

A simple contrast will exhibit the difference. When the Ayatollah Khomeini pronounced Salman Rushdie's novel *The Satanic Verses* to be contrary to the Koran and condemned the author to death, the first reaction was not to name an ecumenical council of clergy to determine whether the "nature" or "essence" of the book is wrong and, if so, whether the Koran authorized such a sentence. For those Muslims who were outraged by the book, especially in England, Khomeini's judgment was legitimate because the ayatollah made it; for Westerners, accepting the principles of a liberal society and free press, a religious leader in one nation and faith has no *authority* to condemn an author from another nation. No evidence could possibly contribute to the dialogue between two such positions: either the ayatollah was considered a valid authority and his statements correct, or his authority to judge was denied along with the validity of his pronouncement. Even Westerners who disliked Rushdie's novel or considered it insulting to Islam did not generally conclude that the ayatollah's proclamation could be judged by *scientific* means. Ultimately, Rushdie went into hiding to avoid being murdered.

Consider, by way of contrast, the reactions to the claim that "cold fusion" had been produced in a laboratory at the University of Utah in 1989. Although some journalists deferred to the scientific credentials of the experimenters (Stanley Pons and Martin Fleischmann), other scientists demanded to see the evidence. Even more important, they wanted to know whether the predicted phenomena could be reproduced in another laboratory. When this proved impossible, the claims for cold fusion tended to evaporate. The evidence, not the authority of the individual scientist, is what counts in modern science. Ultimately, one of the physicists who originally claimed to have demonstrated the effects (Stanley Pons) went into hiding to avoid answering questions about his work.[4]

Philosophers and historians of science debate the extent to which this ideal of the free play of contradictory evidence is actually the basis of scientific theories and controls their acceptance. For some, modern science approximates the norm of a system of hypothesis formation and empirical confirmation. For others, following Sir Karl Popper, scientific knowledge is more modestly defined by "falsifiability"—the willingness, in principle, to submit one's hypotheses and data to test by others. For yet others, the hidden elements of authority and convention qualify this process in a fundamental manner so that science ultimately reflects the norms of a society rather than reality itself. But in any of these interpretations, modern science does not discover absolute, universal truths or rely on argument from the authority of a person or office as the last resort in conflicts over knowledge and truth. An individual's virtue, faith, or power doesn't guarantee scientific validity.

Second, modern science does not claim wisdom or truth about the ends of life but instead focuses on methods of research that lead to the discovery of regularities, or "laws," of nature. Ancient religious and philosophic teachings sought to provide knowledge of the way to live—and die. Modern scientists typically refuse to extend the hypothetical and empirical methods of science to the primary goals of individual or collective life; when they do so, it is often with a guilty conscience.

In assessing the future implications of human sociobiology, John Beckstrom has used an analogy that epitomizes this prevailing view of science: "Biocultural science may eventually serve a function similar to that of an airline ticket office. It may be of little or no help in telling us where we ought to go, but it may help us estimate the costs of getting there and help us to make the journey." [5] Even where a modern scientist claims to have knowledge relevant to human behavior and the public policies or laws regulating it, scientific expertise is focused on the means to be used, not the ends to be sought.

This change is symbolized by the concept of a "law of nature," typically used to describe the regular patterns discovered by the natural sciences. [6] The focus of modern science on regularities, rather than ends or purposes, leads to the primacy of method and of mathematical formalization. In principle, any trained scientist should be capable of testing a newly reported finding. For this to be possible, there must be publicly teachable methods that serve as the basis for the necessary procedures. Hence, when scientific papers present empirical evidence for or against a hypothesis, they generally include a section on the methods used (often specifying the machines used and other details of procedure).

Among these methods, measures of statistical significance are particularly important. Contemporary scientists usually accept without question Hume's claim that the correlation between two events does not prove there is a causal connection between them. Scientific explanations are therefore viewed as only provisional or probable. Tests of statistical significance, such as those introduced by R. A. Fisher, are often crucial: if the findings could have occurred by chance more than five times in one hundred cases, they are usually not considered worthy of acceptance. [7]

Finally, the third negative characteristic defining modern science is social: *it is not normally possible to do science as an isolated individual.* Religious teachers can claim to have received illumination or divine grace in solitude. Ancient sages argued that they had come, through individual effort, to a systematic understanding of nature or to wisdom about human affairs. Modern science, in contrast, is a social system. More and more, articles by multiple authors replace books by an individual theorist. More and more, specialization replaces breadth of expertise. As Ortega put it, modern science is the first form of knowledge in which ordinary and even mediocre specialists can add to theoretical understanding.[8]

This feature of scientific knowing has an important corollary. In modern science, it is not legitimate to form sects or schools closed to all but insiders. For the scientist, there should not be an esoteric or hidden doctrine, accessible only to the elect.[9] Scientific knowledge must, as a matter of principle (albeit not necessarily in practice), be available to all because it must be subjected to test by others who reject the scientist's authority and doubt the reported findings. Training may be necessary to understand the theory or evidence, but scientists must publish their results and subject them to criticism from peers, who may not agree on research methods or theoretical premises. And because a valid theory or hypothesis needs to be presented in a form that can be accessible to others, scientific disciplines develop conventions and theoretical paradigms that often seem to change only when the supporters of an older theory die and are replaced by younger scientists.[10]

The Origins of Modern Science

Where did this system come from? The theories, methods, and institutions we associate with "science" arose at a specific moment in the history of Western civilization. While other cultures have had forms of rational inquiry and technological expertise, only the West developed the self-reinforcing system of scientific research, theoretical controversy, technical invention, and productive know-how that is taken for granted in contemporary industrial societies. We cannot hope to understand our current situation without considering its historical origins.

In the fifteenth century, notably in Italy, artists and thinkers rediscovered the artifacts, techniques, and writings of the ancient Greeks and Romans.[11] The movement we call the Renaissance was, however, more than a simple return to antique ways. On the contrary, this rediscovery of the pagan Mediterranean civilizations was associated with a sense of novelty: in going back to the ancients, there was also a quest for a form of humanism that was without precedent.

Two thinkers epitomize this movement, Leonardo da Vinci (who represents radical innovation in science and technology) and Niccolò Machiavelli (who initiated a transformation in political and social thought). It is especially appropriate to focus on Leonardo and Machiavelli because their friendship—which seems to have escaped the notice of scholars—may well have been a central event in the emergence of modernity.[12]

Although Leonardo is best known as an artist, he thought of painting as "a subtle invention which brings philosophy and subtle speculation to bear on the nature of all forms"—and hence as a "science" providing the basis of understanding nature.[13] It followed that all conventional techniques in art, industry, and thought needed to be reconsidered in the light of a rational quest for order, form, and power. Leonardo's inventiveness, in areas ranging from the design of weapons and military defenses to pigments and artistic effects, is legendary. More relevant, however, is his understanding of what came to be called the scientific method.

Leonardo's *Notebooks* contain the outline of an innovative natural science based on experimental evidence: "First I shall test by experiment before I proceed farther, because my intention is to consult experience first and then with reasoning show why such experience is bound to operate in such a way."[14] Leonardo's science is not, however, merely based on untutored evidence. He saw that technical advances depend on scientific knowledge and that scientific knowledge depends on mathematical proof: "no human investigation can be called true science without passing through mathematical tests."[15] The effect of such a "true science," moreover, is the human ability to control or overcome natural forces that hitherto seemed invincible: Leonardo not only imagined aircraft, submarines, and many other inventions that have been realized only in our lifetime, but he also apparently dreamed of a musical instrument whose sounds would not decay and die out—that is, a perpetual motion machine that would conquer the natural limitation of sound.

Whereas Leonardo represents that approach to natural science and technology we associate with modern life, his young friend Machiavelli epitomizes the way such a humanistic science would relate to human affairs. Like Leonardo, Machiavelli goes back to ancient models—particularly to the ancient Romans—as a way of transcending the limits of Christian humility.[16] Like Leonardo, Machiavelli's return to antiquity is in the name of radical novelty: "I have decided to enter upon a new way, as yet untrodden by anyone else."[17] Like Leonardo, the end is the transformation of human life into a "work of art" under the control of the creative human personality—in politics, the legislator, or "entirely new prince," whose "new modes and orders" can give shape to a more successful regime than has hitherto been possible except by sheer chance.[18] And while Machiavelli does not seem to conceive of a *permanent* conquest of nature, he foresees the possibility that humans "might" be able to "govern" up to "half" of the events often attributed to fortune or historical necessity.[19]

In the sixteenth century, what can be called the modern scientific project took more explicit form in the works of Sir Francis Bacon. While many see in Baconian science little more than the replacement of medieval scholasticism by more empirical methods, his intentions go much farther. Bacon claims that it is possible for humans to use scientific knowledge for the "conquest of nature." To be sure, to conquer natural necessity one must first learn the rules or laws of nature and obey them.[20] But once natural processes are known and controlled, Bacon sees a radical possibility of the "relief of man's estate": by imposing "new natures" on old, the new science can attain the power to remove the limits of wealth and prosperity that have hitherto constrained all human regimes.[21]

In place of the ancient science of the proper *ends*, or perfection of human life, this new science focuses on the efficient and powerful *means* to overcome natural necessity. Bacon himself was not concerned about the ends to which power over nature would be used.[22] For Bacon, happiness will be possible on earth if, and only if, the modern scientific project is organized into an effective scientific and political community.

In the seventeenth and eighteenth centuries, this Baconian project was pursued by a small number of outstanding scientists and thinkers. In the natural sciences, Galileo and Newton applied Leonardo's insight to a new understanding of physics as a system of universal laws governing matter in motion. The medieval image of a universe was full of crystalline spheres, invisible spirits, and innate essences. Instead, Galileo, Newton, and their followers saw a world composed of bodies moving through a void along trajectories described by mathematical curves or functions.

Mathematics, of course, played a critical role in this transformation of ancient natural philosophy into modern physics; whereas the ancients had never solved the intellectual problem of understanding continuous motion, the invention of the calculus by Newton and Liebniz opened the possibility for a science of motion.[23] For the Aristotelians, one could not go far beyond the idea that rest is the natural end of motion: bodies "naturally" fall to earth. Armed with the calculus, one can combine Galileo's concept of inertia—eternal motion as the natural consequence of movement once begun—with the Newtonian concept of gravity as a force of attraction between bodies, thereby emptying the visible world of natural purposes or goals.

Descartes applies this view of nature to human thought itself, with im-

portant implications for the self-consciousness about method that characterizes modern science. From the *Rules for the Direction of the Mind*
to the more famous *Discourse on Method*, Descartes seeks to harness the
skepticism of the ancients to a radically new mode of understanding how
humans can discover the truth.

In all thought, as in mathematics, Descartes sees that humans need a
procedure (or algorithm) such that the mind can divide previously intractable problems into infinitesimally small but "clear and distinct" steps.
Armed with such a method, the confusion and chance of human life can be
reduced to order, predictability, and therewith greater happiness.[24] Such a
conception of natural science could not fail to have political implications.

Hobbes seems to have been one of the first major thinkers to utilize
mathematics as a means of transforming the science of human nature,
though he combined Galileo's view of inertia with Euclidean geometry
rather than with the calculus that, after Newton and Leibniz, came to
characterize modern science. For Hobbes, the new science is based on
hypothetical propositions in predictive form (if x, then y), which must
then be subjected to empirical confirmation.[25] Such propositions depend on
the proper definition of terms, for only definitional exactitude can permit
deductive proofs, whether in geometry or in politics.

Following this procedure, Hobbes concludes that confusion and conflict
in human affairs can be overcome by a scientific understanding of politics.
The basic principle of this new science is that all humans have a "natural right" to preserve themselves, much as moving physical objects have a
natural tendency to continued motion (the "inertia" of Galileo's physics).
Hobbes even suggests that, once developed and generally accepted, the resulting science would make possible an "eternal" commonwealth capable
of controlling the otherwise insuperable chaos and misery of human affairs.[26]

Although Hobbes's teaching repelled many readers, Locke used a similar understanding of natural rights (albeit presented with a gentler rhetoric)
to establish the modern view of representative or constitutional government.[27] Along with the scientific perspective of Newton and Descartes,
Lockean republicanism justified the Glorious Revolution in England and
became the basis of the enlightenment challenge to the ancien régime on
the Continent.[28] After the French Revolution demolished the intellectual
hegemony of the church and the political dominance of inherited status,
modern science could become fully organized as a social system in which
the community of scientific endeavor is at home in liberal or republican
political institutions.

The interaction between politics and science, emphasized in this brief historical account, deserves emphasis. What we think of as modern science presupposes a community of scholars free to engage in research and publication: as long as a Galileo could be forced to a public recantation of scientific theories for political or theological reasons, modern science as we know it could not fully come into existence. The emergence of modernity is symbolized by the transition from the burning of Giordano Bruno as a heretic in 1600 to the celebration of Newton's scientific genius in 1700.[29]

In this sense, we can see a first limitation of the modern scientific project: it cannot exist without a favorable political and cultural context. But perhaps this historical circumstance is irreversible, reflecting a definitive conquest of the necessities that had controlled all prior societies. If so, the fact that modern science and technology are possible only in a secular, progressive political order would be a tautology flowing from the character of human history. To assess the modern scientific project, it is not enough to describe its origins; we must also analyze its methods and character.

The Methods and Character of Modern Science

While other scholars will have their own mode of characterizing modern science, the following traits seem a reasonable description of the scientific enterprise as it is practiced today:

1. The world is *knowable* rather than capricious or marked by miracles whose explanation is impossible without faith in God. While chance or chaos may well occur, such processes are a specific pattern of events that can be represented mathematically and need to be distinguished from religious, mystical, or mythical beliefs.

2. The world is *reasonable but not certain* both in the sense that our knowledge can never be absolute and in the sense that events themselves are often matters of probability rather than rigid determinism.

3. Human understanding of the world is *communicable* to any normal person by both verbal and nonverbal means. When language is used, it is typically necessary to define, if not to create, specialized terms, although mathematical formalizations are often likely to be even more precise than language.

4. Because the "essence" of things is beyond human knowing, the language and mathematical formulas used to describe the world are "constructs" or "representations" to be assessed in terms of *utility*, not proposed as if they were eternal truths. Scientific propositions are therefore most rigorously stated in the form of hypotheses (if x, then y).

5. Methods of discovering the scientific knowledge must be *operational*, including precise specifications of the instruments and procedures needed

to reproduce the discoveries of others. The scientist's personal interest and the desire for fame or economic benefit cannot justify a refusal to share the information needed to test the accuracy of results and falsify the hypotheses based on them.[30]

6. Building on scientific procedures that can be repeated by others, the knowledge available to the scientific community is *cumulative* and hence, at least in principle, *progressive*. New theories and the evidence used to test them build, and *must* build, on work that has gone before.

7. The results of such scientific endeavors are *beneficial* to the human species, most particularly because they give rise to *technological innovation* that can improve human life.

Reflection on this list indicates that, in some respects, modern science is not unique in human intellectual history. In other epochs and cultural contexts, serious thinkers have clearly held that the world is *knowable* rather than capricious or marked by miracles whose explanation is impossible without belief in God and that the world is *reasonable* but not certain. As has been suggested above, there is more than one alternative to modern science; some forms of religious belief entail a rejection of all seven of the premises listed above, and ancient science shares the first two but rejects the last five.

Technology, Desire, and Society

Modern science differs from previous modes of knowing in good part because it continuously generates new technological processes and products that in turn transform social practices. This marriage of theoretical understanding and practical technique did not exist in ancient science. How is it associated with the "value-free" approach of today's scientists?

From its inception, as can be seen in the *Notebooks* of Leonardo, the modern scientific temper was associated with the construction of machines and devices that enhance human power and make life easier. Like other inventors in the fifteenth and sixteenth centuries, Leonardo suggested improvements in military technology as an advisor to rulers; unlike his contemporaries, he imagined machines like submarines and airplanes that were not developed until the twentieth century. Moreover, Leonardo's projects included improvements in urban and domestic life: town planning, domestic architecture, controlled water supplies, machines for minting currency, and numerous other mechanical devices.[31]

Since Leonardo's time, extensions of scientific knowledge have produced new technologies that make human life easier, and these techniques

in turn have led to further scientific advances. Ultimately, it is this reciprocal relationship between science and techology that distinguishes modern
science from that of the ancients.[32]

Fruits of modern science and technology with a dazzling variety of
uses seem to have a single underlying principle. Virtually all inventions
reduce the delay between the first desire for a pleasure or experience and
its satisfaction. Radio and television give us immediate information from
everywhere in the world. Telephones allow us to speak with almost anyone instantaneously. For those who can afford them, automobiles, trains,
and especially airplanes make it possible to get from one place on earth to
another in no more than forty-eight hours, and usually less. Refrigerators
allow us to eat virtually anything we wish at any time. With a home VCR,
one need no longer go to the theater to watch a play, movie, or opera. With
a fax or electronic mail, one need no longer wait for the postman to deliver
a message. Hot water runs from the tap, and depending on climate, central
heating keeps us warm or air conditioning, cool.

These technological marvels, most of which were predicted by Leonardo
but realized only in the past century, have a further common feature: the
instrument of satisfaction is divorced from judgments of whether that satisfaction is itself "good" or "bad." The equipment used to transmit radio,
television, telephone, fax, or computer messages is indifferent to their content; a refrigerator preserves codfish or caviar; the VCR, compact disc,
or audio tape can play a Beethoven symphony, sentimental ballads, or
pornographic pictures and music.

It follows that many technological advances can be neutral to the ends
or purposes to which they are put. New methods of birth control can be
used to encourage responsible family planning or sexual promiscuity. The
phenomenon is evident in public matters as well: nuclear weapons were developed by both constitutional democracies and Communist dictatorships.
The inventions that mark our civilization are means to satisfy desire, but
they seem equally open to what traditionally was described as the virtue or
vice of the user or the user's purposes.[33]

In this sense, modern science and technology are in principle egalitarian
or democratic.[34] Our machines make it possible for almost everyone to do
things that were once the preserve of a gifted or wealthy few. In today's
highly industrialized society, even many on welfare have conveniences that
were available only to high nobility during the Middle Ages: cold (if not
hot) running water and indoor toilets, music or entertainment at a whim,
varied food (often available without reference to season), covered convey-
ences for travel, shoes and waterproof clothing. One doesn't need to be
musical to hear music (any radio or tape player will do), to be wealthy to

possess paintings by the great masters (reproductions are everywhere), nor to be physically strong to climb mountains for sport (as long as resorts are equipped with chair lifts or tramways).

All pleasures and experiences are equalized. In such a world, the market economy seems the ideal mechanism for allocating goods and services. Neither rulers nor priests can claim to determine what is good for the consumer. Rather, the laws of supply and demand leave the issue to the individual's choice among competing products or experiences: "You pays your money and you takes your choice." Centrally planned economies may be justified in time of war or during a period of forced investment, but their current demise throughout the world seems to stem from the contradiction between the principles of modern technology and centralized control over the distribution of its fruits.

This aspect of contemporary technology mirrors a theoretical premise of modern politics. Whereas ancient political philosophers stressed virtue and duty, moderns have come to focus on individual "natural" or civil rights, stressing freedom as the basic principle of social life.[35] The American Declaration of Independence speaks of the "right to life, liberty, and the pursuit of happiness"; it does not define what happiness might be nor how one should pursue it. Hobbes saw this clearly when he sought to establish a scientific understanding of the process of human thought and desire on the grounds that a science of goals or purposes was impossible.[36]

Although our scientific and technological system seems neutral with regard to values or goals, its fruits are not without social or moral implications. On the contrary, many of the inventions of our time have hidden consequences for society. Anyone can see that the automobile has spawned the suburbs, extending the distances of daily travel to and from work and transforming the character of cities. But not all of these consequences are obvious, and some (particularly when the relationship between a technological innovation and its social consequences are obscure) entail moral controversy.

The telephone, and especially the telephone switchboard, is a good example of an invention that has had hidden as well as visible effects on social organization. The changes it has produced in patterns of communication are obvious. Less so have been its effects on architecture and urban design. Yet at least one scholar has made a convincing case that it is the telephone, even more than the elevator, that was essential for the development of the high-rise building. Without telephone systems, so the argument goes, those on higher floors would find it impossible to communicate without constantly sending messengers, thereby overtaxing systems of vertical movement.[37]

Modern science and technology have thus unleashed continued and un-predicted social transformations. Even if these devices did not seem value-neutral toward their uses and users, this process would necessarily be un-settling to those who prefer established patterns of life. The concept of "progress," embedded in the optimistic projections of the development of ever new techniques, places a favorable value on change. In fact, however, many of the social transformations due to scientific and technical inno-vation have been viewed as morally wicked or politically undesirable by substantial portions of the community. And since modern science does not pretend to establish the criteria for using its technological offspring, the result is the sense of malaise associated with the "crisis" of our time.

The Moral Consequences of Technological Change

It might seem at first that criticism of science and technology is the work of "conservatives," who are opposed to change as a matter of principle. Such an assumption is easily disproved: violent hostility to nuclear tech-nology (and nuclear weapons), which fueled the peace movement during the 1960s, was primarily supported by the "progressive" left; in most indus-trial societies, it has been the so-called conservatives who view nuclear power most favorably. Whether a specific innovation is attacked by the conservatives, by the liberals, or by environmentalists who cannot always be classified in terms of left and right seems to depend on circumstance rather than principle.

Criticism of the fruits of modern science and technology does not usually come from any one political principle or ideology; instead, when an inno-vation or scientific theory contradicts established interests or beliefs, those involved attack the innovations on moral grounds. Galileo was forced to recant by the Inquisition, just as biologists who accepted Mendelian genet-ics were silenced by Lysenko and the Stalinists. There are conservatives and Catholics who oppose birth control and abortion but favor nuclear power, and liberals who support the medical technologies that have revolution-ized family planning while attacking industrial processes that threaten the environment.

When a technology has obvious effects on social behavior, these politi-cal and moral debates are relatively easy to understand even when they are difficult to resolve. That nuclear power creates danger to life and health has been evident from its inception and inescapable after Chernobyl; the contentious question concerns "risk assessment" (the extent to which tech-nological controls have been effective). That industrial products can pollute the environment is hardly mysterious, though the facts are often bitterly debated in specific cases (witness the controversies over acid rain and global warming).

When this happens, scientists and technicians are often found on opposed sides of the controversy, generating public skepticism about scientific evidence. Little matter that in the domain of modern science such controversy is not merely legitimate but essential. As we have seen, the hallmark of the contemporary approach to knowledge is the challenge of all proposed explanations by those who claim to have found contrary evidence or an alternative theory. Not understanding the scientific method, the public and its political leaders seek quick answers and, when they are not forthcoming, blame science and technology itself.[38]

Even more unsettling, however, are political and moral issues that result from unsuspected or indirect effects of science and technology. In these cases, an unpalatable practice is often attacked as if it emerged as a result of misguided beliefs or moral obtuseness. The most careful analysis may be needed to show how a technological innovation has generated the problem. Lacking that understanding, the controversial situation persists despite vigorous attempts to eradicate it. The result is that many feel unable to cope with the social and moral world around them.

An example will help. At the turn of the twentieth century, divorce was reprehensible for many, if not most, families in Western Europe and North America. Catholics were far from alone in condemning the practice. It was not uncommon for those who were divorced to be shunned, especially in "good company." As recently as 1952, Adlai Stevenson's political career was compromised by his divorce. Given the general condemnation of divorce, why has marriage become so fragile? The answer requires an understanding of the way modern technology creates a new environment, akin to a transformation of the ecosystem.

The refrigerator is a divorce-making machine. This assertion often brings puzzled expressions to the faces of those who hear it. Yet there is much reason to believe that it is the development of the refrigerator and other consumer durables like it that lies at the root of the radical increase in the frequency of divorce. How could this be? To answer, one has to consider why males and females of other animal species form lasting pair bonds. A brief excursion into the biological discipline known as behavioral ecology is therefore necessary.

Some animals are monogamous: one male mates with one female for life. Other animals are polygamous (one male mates with a number of females), a few are polyandrous (one female mates with a number of males, usually brothers). Sometimes sexual intercourse is limited to brief consortships, without any lasting bonds like those of the human family. Leaving to one side the complex reproductive systems of social insects, a wide variety of reproductive and family patterns can be observed among birds and mammals.[39] Similar variations can, of course, be observed in humans.

Among other animal species, the reason for mating patterns can be traced to the physical and social environment. As a general rule, where the benefits of a lasting pair bond are greater than the costs, animals form monogamous pairs; among primates, a good example is the gibbon. Sometimes factors like food scarcity or predation may lead to polygamous mating, with one male monopolizing a number of females, as is seen in the hamadryas baboon. Where an animal can rely on food supplies that are abundant and available in small packets easily consumed by the individual and where no predators threaten, we often see short consortships or promiscuous mating, as among the orangutan (for whom there are no lasting pair bonds).[40]

There is no reason that such environmental factors should be without influence on human mating and marriage systems. On the contrary, the evidence shows that similar considerations of cost and benefit apply to humans.[41] From the perspective of a cultural anthropologist, economist, or sociologist, as well as for the evolutionary biologist, varied practices associated with kinship, marriage, and child rearing are questions of "objective" scientific knowledge, not of moral condemnation.[42] And from such a point of view, the refrigerator is indeed a "divorce-making machine."

Before refrigeration and other machines invaded the twentieth-century home, managing the household and rearing children was a very time-consuming activity. To purchase and cook food, clean clothing, and take care of children was a full-time job. A husband who no longer loved his wife was well-advised to take a mistress rather than get a divorce; a single man would not have time to work, shop, cook, wash, and clean. Similarly, his wife often found it easier to have an affair than to break up the marriage; a divorcée without alimony had few ways of earning income while taking care of her house and children.

The refrigerator changed all that. Now a single person, like an orangutan, lives in an environment in which food supplies are abundant and easily harvested by the individual. Males and females can forage independently. Consortships last if convenient, but their durability is not reinforced by the material environment, with the result that cultural expectations have changed. We've become accustomed to the fact that about half of all marriages in the United States end in divorce.

Some may contest this explanation of the fragility of marriage today. Whatever the actual causes, the issue points to the problem. Our tendency is to hold responsible the individuals who divorce; when the practice became highly frequent, social critics were tempted to see in it moral decay. Yet it is at least possible that the true culprit was the refrigerator along with the washer, dryer, vacuum cleaner, and other consumer durables, which the same social critics may well have praised as symbols of the progress of industrial society.

Little wonder, then, that our society faces a sense of uneasiness. As the global balance of power changes, who will control nuclear weapons and the sophisticated techniques for delivering so-called conventional arms? As biologists discover exactly how genes work and unravel the mysteries of the human brain, new technologies arise for the manipulation of behavior. Even at a trivial level, we face uncomfortable changes: miniaturized computers permit our refrigerators to speak, telling us that the ice-cube tray is empty.

To understand and respond to this situation, we need to reconsider the foundations of modern science and technology. Specific issues cannot be solved one at a time, for the basis of the problem is deeper. Why is our approach to scientific knowledge and its application to human life divorced from the reasoned discussion of the ends or purposes of human life? To answer this question, we need to consider the origins and limits of what is known as scientific value relativism.

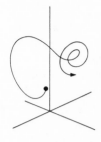

3. Ancient Science:
Time, Form, and Knowledge

Why Consider Ancient Science?

According to the methods of modern science, all scientific propositions are merely provisional hypotheses that need to be tested by the relevant evidence. Why, then, have most contemporary scientists failed to question the conception of a gulf between fact and value, *Is* and *Ought*, means and ends? And why has society at large accepted this divorce between scientific knowledge and moral or political judgment?

If the scientific endeavor is to be true to its principles, the claim that pure science is (and ought to be) value-free cannot be placed above critical inquiry. The same, of course, needs to be said for the attack on all knowledge as value-laden or culturally relative. Both forms of relativism need to be reexamined, not only in the light of recent discoveries about nature and human society but also with regard to the difference between ancient and modern science.

Most contemporaries take it for granted that science has progressed since Greek and Roman antiquity. Plato and Aristotle, not to mention Heraclitus, Epicurus, and Zeno, are viewed as philosophers, rather than as scientists in the strict sense. The evidence suggests that this assumption needs to be reconsidered. Perhaps we have made a mistake in equating technological power with knowledge and, in so doing, have been led to questionable assumptions about how science relates to ethics.

In a number of disciplines, the latest theories and empirical findings

point to crucial limitations in scientific assumptions that were conventionally accepted a generation ago. In the natural sciences, and especially in
mathematics, it is becoming apparent that the modern understanding of
continuity in time and space is not always adequate. Even the notion of
"laws of nature" (now the typical way of describing scientific knowledge)
is open to question. The focus on the form and character of things—which
was central to ancient science—no longer seems absurd or unscientific.

Critics of contemporary relativism have emphasized the moral dimension of modern science and society. Paradoxically, the crucial issue may lie
in mathematics and the different conceptions of time, form, and knowledge in modern and ancient science. Today, most scientists, especially in
the social sciences, seem to have adopted a mathematics whose power is
unquestionable but whose scope is surprisingly limited. An exploration
of these limits is essential if we are to answer the question of values and
relativism.

For the ancients, mathematics was the highest science—a realm of pure
intelligibility. In modern science, mathematics became an instrumental
method of measurement and verification—a means to discover and control
nature. Ancient mathematics, epitomized by Euclid, was oriented toward
geometry and the study of form. Modern mathematics is symbolized by
algebra and the calculus of Newton and Descartes. Before exploring the
implications of this distinction, we must understand it.*

Mathematical Thinking: Time and Form in Ancient and Modern Science

There can be little doubt that a major reason modern science differs from
that of the ancients is the transformation in mathematics that took place
between 1500 and 1700. At the beginning of this period, as in antiquity,
there was no precise mathematical way to represent continuous motion:
as far as we know, Zeno's paradox was as insoluble for Machiavelli's contemporaries as it was for thinkers in Plato's time. By the beginning of
the eighteenth century, the calculus had been invented—to some extent
independently—by Newton and Leibniz. Before the Renaissance, and indeed as late as Hobbes's publication of *Leviathan* in 1651, the model of a
mathematical demonstration was Euclidian geometry.[1] Once armed with

---

*At this point in the argument, let the reader be warned: although the implications of
mathematics are an essential element of serious philosophic discourse, I am not a specialist
in the disciplines about to be discussed. While I cannot pretend to a definitive interpretation,
there is no alternative to considering the relationship between mathematics and the understanding of inanimate as well as animate nature from a perspective comprehensible to the
nonspecialist.

the calculus, however, scientists could claim a new kind of mathematical proof based on algebraic formulations.

The new mathematical models of nature are sometimes said to have originated with the need to predict the trajectories of the new and more powerful cannon produced in the Italian Renaissance. Scientific innovation was furthered by Galileo's radical transformation of the understanding of motion. For the ancients, bodies were "naturally" at rest, as could be observed by the tendency of motion to cease as bodies fell to the ground or stopped rolling. Galileo emptied the universe of the objects that might impede motion and in so doing introduced the concept of inertia.[2] Appearances to the contrary, bodies move eternally unless something interferes with this motion, either by stopping it entirely (as the earth does when a ball falls to the ground) or by influencing its trajectory (as happens to the artillery shell before it hits its target).

Although ancient astronomy had developed extraordinarily acute devices for observing the heavenly bodies and their motions,[3] the resulting patterns were understood in terms of geometric shapes—most notably, of course, the circular orbits of the Ptolemaic theory of the solar system. If the observations did not fit the geometric form, it was necessary to multiply forms, as in the Ptolemaic system of epicycles (circles on circles). It was not an equation but rather a geometric form that could describe the motion of a planet or the moon. At its extreme, as among the Pythagoreans, all nature was said to be understandable solely in terms of these geometric shapes or forms.

Galileo's concept of inertia meant that observed motion will most often be the product or result of several interacting forces (or to use the contemporary term, a vector). A solution to continuous measurement of motion, which to the ancients was interesting (as is evidenced by Zeno's paradox), became absolutely essential. If two different forces interact, it is necessary to be able to calculate what the motion would have been had each acted alone (the simplest case being inertial motion in a straight line) and then to weight the two forces to estimate the result. But this means that in addition to developing measures of continuous motion, it was necessary to be able to predict the location of a body given its original movement and the specified forces acting upon it. In short, equations were needed to replace the classic emphasis on geometric forms or patterns.[4]

Ultimately, the model of the universe was utterly transformed: the Ptolemaic worldview of circular orbits with the earth at the center was replaced by empty Newtonian space in which points move along trajectories. Mass, velocity, distance, acceleration, gravitational attraction became terms in equations, permitting a description of the path of motion. And of course, knowing the velocity and the distance of a body's motion, one can immediately calculate the time elapsed during the movement.

To achieve this goal, variables must be *continuous*. Calculus makes no sense unless each foot or mile of distance and each minute or year of time is absolutely equal to all others. Texture and particularity are replaced by smoothness; in contemporary statistics, the graph of a regression equation creates the *smoothest* line or curve joining the observed points. Space becomes an empty multidimensional space within which bodies interact and move along linear pathways or vectors (an image of the cosmos that has only recently come into question); similarly, time becomes a smooth continuity of instants aligned in linear sequence.

Natural processes could now be represented by equations, of which the most famous is probably $e = mc^2$. Until recently, such equations have had two interesting properties. First, they were typically linear or continuous, that is, equations in which changes in an independent variable produce proportional changes in the dependent variable without discontinuities. And second, results were deterministic; that is, knowledge of all independent variables in the equation permits one to compute (or predict) the dependent variable. Leaving aside quadratics, normally such equations have a single "solution," which is determined by the value of each variable and the relationship between variables.

This manner of thinking is taken for granted today, but it was an astounding innovation in human intellectual history. Ad hoc mathematical computations could be replaced by a formal or general algorithm in which qualities and motions are represented by abstract symbols. The calculating machines of Pascal and Leibniz converted the laborious effort of arithmetical computation into a simple action, for which one needed only to enter the specific numbers to be computed; the mathematical equations that became the core of scientific reasoning, especially in physics, used algebraic representations of movement and change in the same way. Time and place as experienced naïvely do not have this character. Modern science replaces the experiential discontinuities of life with the quest for the precise algorithm or equation that allows one to *predict* the outcome—if possible, with certainty.

The algorithm is thus a rule or set of rules, instructing the user on the way to compute the interaction of attributes that have been defined as variables. Simple phenomena—the rate of movement of a ball down an inclined plane—can be described and predicted by a single equation. Similarly, an equation can often generate a geometric shape. More complex events may require a system of equations. Ultimately, the mystery and unpredictability of countless natural phenomena, from the ocean's tides to the phases of the moon, were explained rationally and predicted with exactitude by such equations. Little wonder that the transformation effectuated by mathematical representation came to dominate the concept of natural science, leaving those who are ill at ease with algebra the choice between poetry and politics.

The ancients had a quite different model of mathematics as the foundation of scientific reason. Whereas modern science rests on algebraic equations, ancient science rests on geometry and the deductive proof, moving from axioms or definitions to a step-by-step *demonstration* that a given form or pattern must have a specific property because any alternative conclusion contradicts at least one of the original premises. Such a demonstration rests on the assumption that the definitions, axioms, and premises hold true throughout the entire (timeless) examination of the proposition being studied. A conclusion is demonstrable when any other outcome necessarily contradicts what has already been given, reducing objections to absurdity (the reductio ad absurdum).

Such a mathematics does not produce predictions or calculations. Rather, it explores relationships. Space and time are not uniform and continuous; rather, the universe is composed of distinct and highly differentiated forms, each of which has a character that can be understood rationally only by careful analysis and deductive proof of those qualities without which the thing would not be itself. The focus is not on time as a continuous variable embedded in the equations describing uniform movement through empty space. Instead, the focus is on *form* as a characteristic shape or pattern that retains its basic or essential features regardless of time and place. As a result, for most ancient thinkers, understanding of the nature of a thing—or of nature more generally—cannot be entirely reduced to the process of its coming into being.[5]

Although the ancients knew that some things or measurements in nature are continuous, they resisted the modern notion that ratios and numbers are in principle homogeneous. For Euclidean geometers, the set "⅓ . . . ⁴⁄₃ . . . ⁵⁄₃ . . . ⁶⁄₃" does not form a continuous series: a whole number (1 or 2—or in this instance, ⅓ and ⁶⁄₃) is "commensurate" and thus different in kind from a fraction or ratio (*logos*) that is "incommensurate" (⁴⁄₃ or ⁵⁄₃). Today, since all notations in modern mathematics are ultimately operators or measures, it is a convention (i.e., a question of utility or convenience) whether the series is noted "1, ⁴⁄₃, ⁵⁄₃, 2" or "1, 1.333, 1.666, 2." Not so for the science of antiquity, in which the heterogeneity of "natural kinds" lies at the root of all mathematical thinking.[6]

Chaos Theory and Nonlinearity

What difference does it make that modern science is based on mathematical assumptions that were not shared by ancient scientists from Thales to Aristotle, Plato, Theophrastus, or Epicurus? The linear equations of algebra seem obviously more general than the static forms of Euclidean geometry: on the one hand, the rules or algorithms for generating these forms can be specified in algebraic formulations; on the other, the resulting ability to analyze time as well as form makes the modern solution far more general than that of the Greeks.

The conventional notion of scientifically discoverable laws of nature rests on these assumptions. In principle, a formal mathematical model that can explain the past should predict the probable outcome in every circumstance. In a determinate physical system, once given the initial conditions, the same equation—or a set of differential equations—should always lead to the same predicted outcome. As a leading philosopher of science put it: "[G]eneral laws have quite analogous functions in history and in the natural sciences. . . . By a general law, we shall here understand a statement of universal conditional form which is capable of being confirmed or disconfirmed by suitable empirical finding . . . a universal hypothesis may be assumed to assert a regularity of the following type: In every case where an event of a specified kind C occurs at a certain place and time, an event of a specified kind E will occur at a place and time which is related in a specified manner to the place and time of the occurrence of the first event."[7] But can it be assumed that natural phenomena are well described by laws of this kind?

Among the most extraordinary developments in recent years has been the emergence of a new mathematical approach, which has come to be known as chaos theory and fractal geometry.[8] More technically, chaos theory concerns *nonlinear* dynamic systems rather than linear ones; that is, determinist systems in which outcomes are not rigidly predictable so that, if a process is repeated, the *same* original conditions (at least within the limits of humanly accessible measurement) can give rise to *different* outcomes. Fractal geometry concerns patterns (often associated with chaos in this sense) differing from the dimensions of everyday perception. These features of chaotic systems need to be distinguished from the indeterminacy associated with quantum mechanics in physics, although both challenge central assumptions of those philosophers of science for whom the perspective of the ancients had been definitively surpassed since the Renaissance.

Chaos theory implies that predictability—and hence the concept that science entails the explanation of events by means of general "laws" of nature—is a special or limited case rather than the defining property of all scientific knowledge. In chaotic systems, undetectable differences in the initial conditions, below the level of resolution entailed in specifying "variables" in the equations or laws that describe the system, can change the outcome in unpredictable ways. Often called the "butterfly effect" (symbolized by the possibility that the motions of a butterfly in the Southern Hemisphere might alter the weather in the Northern Hemisphere), this means that the same values of the variables of one or more simultaneous equations cannot predict the same outcomes.

The modern concept of numbers as homogeneous operators illustrates why the problem of unpredictability in nonlinear systems is of the greatest importance. If numeration is continuous, it is a matter of convention and convenience whether a decimal is rounded to two, four, or more places; 3.14, 3.1416, and 3.14159 are equivalent representations of the value of π (the ratio of the circumference of a circle to its diameter). Actually, of course, 3.14 means 3.14000, which is *less than* 3.14159; similarly, 3.1416 equals 3.1460, which is *more than* 3.14159. In nonlinear dynamic systems, even if determinate, these different values may predict different outcomes or trajectories. As a result, the mathematical assumptions implicit in the calculus and in modern science generally can create artifacts of measurement that make it difficult, if not impossible, to understand many natural and social phenomena.

Quantum mechanics and the introduction of indeterminacy or chance provide an additional challenge to the worldview conventionally associated with modern science. Consider the following summary of an article entitled "Quantum Philosophy" in a recent issue of *Scientific American*: "The deeper physicists inquire into the mysterious world of quantum theory, the stranger it gets. New experiments continue to challenge the common notion of reality. Photons, neutrons, even objects large enough to be seen, lack form until they are observed. Observation can alter the outcome of experiments that have already occurred; measuring one entity can influence another far away."[9]

For the Vienna Circle and philosophers of science responsible for the doctrine of a gulf between *Is* and *Ought*, a science of facts differs from theories of value precisely because of the former makes possible empirical predictions that, in principle if not in practice, take the form of predictive laws. In such a science, postdiction (explanation after the fact, as in the study of history) must be formally similar if not identical to prediction.[10] The difficulty with this view is not (as many humanists and social scientists long argued) that human history is different from natural phenomena;

rather, it seems that even many natural phenomena do not correspond to the picture implied in the notion of linear or predictable "laws of nature."[11]

The combination of chaos theory (determinist systems that are non-linear) with quantum mechanics and Heisenberg's uncertainty principle (phenomena that are essentially indeterminate) means that the conventional notion of laws of nature is at best an approximation. Both the unobservable properties of a system and the fact of observing it can influence the outcome. Prediction (specifying outcomes in advance) and postdiction (defining the causes given the known outcome) are therefore in principle different. One's point of view matters as much in physics as in politics.

A finding of chaos theory helps to explain why the conventional view of scientific laws of nature needs to be placed in a broader context. In many nonlinear systems, there is a domain of linearity such that, for some values in the system of equations, one and only one outcome occurs. Beyond this point, even in determinist systems, multiple but unpredictable outcomes (apparent chaos) or radical changes in the system (apparent catastrophe) can be the consequence. Hence, it is not claimed that the linear assumptions characteristic of modern science are false but merely that they describe only a partial domain of more complex systems.

A well-known instance of such a nonlinear determinist system occurs in population genetics, where equations that were first explored by Robert May gave rise to the concept of bifurcations. Consider the graph in Figure 3.1, which plots the rate of population growth of an animal species on one axis and its population size on the other. Below a given rate of population growth, not enough individuals are born, and the species becomes extinct: the population becomes zero. After that point, increases in rate of population growth produce linear increases in population size; consistent with common sense, the more offspring that are born, the larger the total population. At a point, however, May noted an unusual result from his mathematical simulations: further increases in rates of population growth lead to two different population sizes, one increased and one decreased (the specific outcome being impossible to predict from the initial values in the equations).

Called a bifurcation, this property should not be confused with the butterfly effect; rather, it is an additional property of nonlinear systems, since after one bifurcation there may be two quite determinate outcomes, and the only uncertainty is which of the two will occur in any given instance. Such nonlinear equations can have further bifurcations (making possible four or eight outcomes at a specific level of the independent variable)—or, beyond a point, "chaos" in the sense of apparently indeterminate outcomes. One must, however, insist on the word *apparently*: when

FIGURE 3.1

Bifurcation, Nonlinearity and Chaos

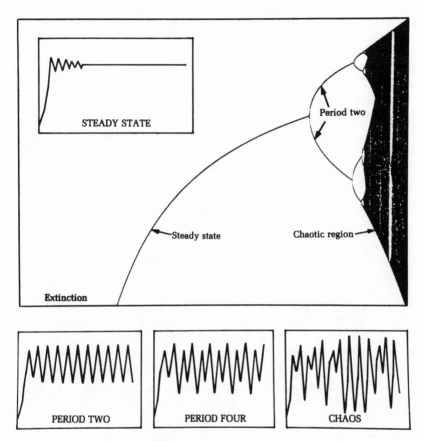

Diagram derived from Robert May's model of population. Vertical axis is population size; horizontal axis is rate of population growth (tendency to "boom and bust"). Reproduced by permission. Source: Gleick, *Chaos*, p. 71. Original legend:

PERIOD DOUBLINGS AND CHAOS. Instead of using individual diagrams to show the behavior of populations with different degrees of fertility, Robert May and other scientists used a "bifurcation diagram" to assemble all the information into a single picture.

The diagram shows how changes in one parameter—in this case, a wildlife population's "boom-and-bustiness"—would change the ultimate behavior of this simple system. Values of the parameter are represented from left to right; the final population is plotted on the vertical axis. In a sense, turning up the parameter value means driving a system harder, increasing its nonlinearity.

Where the parameter is low (*left*), the population becomes extinct. As the parameter rises (*center*), so does the equilibrium level of the population. Then, as the parameter rises further, the equilibrium splits in two, just as turning up the heat in a convecting fluid causes an instability to set in; the population begins to alternate between two different levels. The splittings, or bifurcations, come faster and faster. Then the system turns chaotic (*right*), and the population visits infinitely many different values.

the graphic space describing regions of chaos is magnified or explored more precisely, it is found to contain miniature regions of linearity and bifurcation, replicating the larger picture on a smaller scale.

This discovery of self-similarity is characteristic of another aspect of the new mathematics associated with fractal geometry. Often the repetition of the same operation gives rise to a form (like the Mandelbrot set) that contains within itself smaller patterns of the whole. Few contemporaries have realized, however, that this recurrence of form or pattern at different spatial scales resembles the Aristotelian view of measure.[12]

Linearity and predictability exist in nature, but they do not seem to be the exclusive character of natural processes. Indeed, if one pursues chaos theory further, as in the more recent study of the phenomena of turbulence, form, and structure, it becomes apparent that the principal insights of this mathematical approach involve a transformation of the way scientists look at time and space. In fields as diverse as engineering, paleontology, and economics, the result is a challenge to the seventeenth-century view of universal and predictive laws of nature.

The Rediscovery of Form and the Limits of Modern Science

For the early moderns who developed the calculus, algebra had the immense advantage of making it possible to study time as a continuous variable. This meant that scientific equations could use time as a linear variable, as in Figure 3.2 (*a* and *b*), in which each instant is like every other instant. Once this assumption has been made, one can trace the value of a variable across time, using algebraic formulations to study system dynamics. By challenging the basic assumption on which this mode of analysis rests, chaos theory suggests the utility of focusing on forms or shapes rather than on time.

The difference is clear if one compares Figure 3.2 with Figure 3.3. If the value of a continuous variable (e.g., the position of a body in space) is plotted over time and the system in which it occurs comes to an equilibrium, the line becomes horizontal (Fig. 3.2*a*); if not and the variable oscillates, it forms a jagged line as the body moves from one place to another and back again (Fig. 3.2*b*). One can imagine Figures 3.3*a* and 3.3*b* as looking at the same graphs from a different point of view: the plane at right angles to that of Figures 3.2*a* and 3.2*b*. Hence, Figure 3.3*a* can be seen as looking back along the straight line at the right of Figure 3.2*a* toward the beginning of the oscillation. From this point of view, time is the distance from the viewer's eye—but it is not measured nor measurable

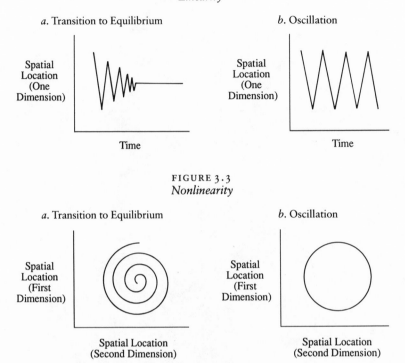

FIGURE 3.2
Linearity

a. Transition to Equilibrium

Spatial
Location
(One
Dimension)

Time

b. Oscillation

Spatial
Location
(One
Dimension)

Time

FIGURE 3.3
Nonlinearity

a. Transition to Equilibrium

Spatial
Location
(First
Dimension)

Spatial Location
(Second Dimension)

b. Oscillation

Spatial
Location
(First
Dimension)

Spatial Location
(Second Dimension)

(and need not be linear); instead, one is able to measure the location of the body in two dimensions, noting that what looked like an oscillation in only one dimension is actually a stable pattern in two dimensions.

When the perspective of Figure 3.3 is used to study systems that do not settle to a fixed equilibrium point or steady state, it gives rise to something called the Lorenz Attractor (Fig. 3.4). Such curves make it possible to study apparently unstable systems in terms of the forms or shapes of movement. Although one may not be able to predict the state of the system at a given time and place, movement is not random. Rather, motion is nonlinear in the time–space manifold, forming a pattern as if drawn to an invisible "strange attractor" (the central holes in Figure 3.4).

One could develop these ideas further by showing how they relate to the rediscovery of form as the basis of scientific understanding of many phenomena. This aspect of chaos theory or nonlinear mathematics is be-

FIGURE 3.4
The Lorenz Attractor

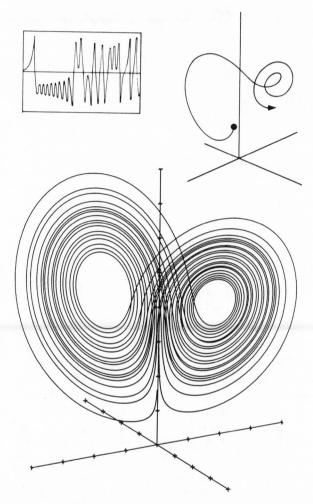

Source: Gleick, *Chaos*, p. 28. Reproduced by permission. "This magical image, resembling an owl's mask or butterfly's wings, became an emblem for the early explorers of chaos. . . . Traditionally, the changing values of any one variable could be displayed in a so-called time series (top). To show the changing relationships among three variables required a different technique. At any instant in time, the three variables fix the location of a point in three-dimensional space; as the system changes, the motion of the point represents the continuously changing variables [upper right]. Because the system never exactly repeats itself, the trajectory never intersects itself. Instead it loops around and around forever. Motion on the attractor is abstract, but it conveys the flavor of the motion of the real system. For example, the crossover from one wing of the attractor to the other corresponds to a reversal in the direction of the spin of the waterwheel or convecting fluid." (Gleick, *Chaos*, p. 29.)

coming well known from the Mandelbrot set and the often extraordinary images that it generates.[13] Such mathematics of form involve the study of recursive systems, in which the state of the system at moment C depends *only* on its state at moment B, and its state at moment B depends *only* on its state at moment A. Whereas the linear systems predict the state of the system at moment C from the state at moment A and the difference in time between A and C, this is no longer possible in nonlinear systems (as should be evident if one considers the second or third bifurcation in Figure 3.1).

Of particular importance is the possibility that perception and recognition of familiar stimuli by the human brain are made possible by patterns of neuronal activity that can be described only by chaos theory. According to a new model of sensory perception, the spontaneous firing of sensory networks produces a chaotic pattern, whereas the discrimination of sensory cues (such as a familiar face or a known odor) elicits bursts of response in a nerve cell assembly that represents a "strange attractor" rather than a linear or determinate effect of stimulus on response.[14] From a philosophic perspective as well, it has been shown that the something akin to a strange attractor is a more general foundation for an evolutionary epistemology.[15]

It is not possible to explore more fully here the implications of the changes involved if one approaches natural phenomena from the perspective of chaos theory and fractal geometry. It takes little reflection, however, to see that nonlinear systems lead to a view of the primacy of form akin to that of ancient mathematics and science. A very simple example of this surprising point is the difference between a modern, algebraic formulation of the equations of population genetics and the description of the same phenomena by Plato.

For a modern biologist, population genetics rests on a logistic equation of the form:

$$\frac{dN}{dt} = \frac{rN(K-N)}{(N)}$$

where N = number of organisms, K = carrying capacity of environment, r = reproduction rate, and t = time.

Such equations traditionally have assumed, as noted above, that the variables are linear and continuous.[16] But in a living population, this cannot be. The actual population at time C can depend only on the population at time B; any unpredicted events that have modified the population between times A and B can have irreversible effects on time C, causing the predictions to deviate from those that could have been made at time A. As a result, the assumption of strict linearity does not hold: events at some times may produce large perturbations, whereas the same event at another moment might not do so. In the evolution of living forms, for example, Stephen Jay Gould argues that if the same process were to be repeated a number of times, each occurrence would lead to different bodily structures and different mixtures of flora and fauna.[17]

The conventional view of population genetics and evolution assumes slow change, leading to steady-state equilibria that are predictable and determinate. The perspective of chaos theory suggests that this is a special case which, while it sometimes occurs, is fundamentally misleading to view as a universal or general law; if anything, at the most critical moments in evolution or history, time is not linear and continuous but effectively discontinuous. Whether a particular evolutionary sequence reflects punctuated equilibria or gradual change is therefore an empirical matter, not a question of abstract theorizing.[18]

In many scientific areas, therefore, mathematical models need to be based on recursive functions whose temporal linearity is no longer assumed as a postulate. This discovery leads from modern algebra back to the ancients and to formal rather than algebraic relationships. For example, consider population genetics. In Plato's *Republic*, the rules of nature governing population are represented by a geometric formula: when stable, the population follows the "round of circles" as in Figure 3.3b; when disturbed, the results are chaotic.[19] Similarly, in many areas of contemporary biology, fractal geometry has appeal as a way of understanding organic form, ecological niche structure, and other nonlinear phenomena.

Linear models, in which time is taken as a continuous variable, produce even more serious difficulties in modern physics. As both Stephen Hawking and Roger Penrose have shown, new models of the cosmos challenge the traditional views of nature that were inherited from the physics of Galileo and Newton. Hence Penrose concludes, as had Heisenberg a generation earlier, that it is necessary to go back to Platonic mathematics and physics.[20] Form or being seems to be prior to temporal change or becoming. While this perspective contrasts sharply with the general tendency to dismiss Plato as an unscientific idealist, it is necessary to look beyond the conventional interpretations of Plato's dialogues to discover the content of their actual teaching; when one does so, it is astonishing to see the superiority of Platonic science to that of moderns like Hobbes, Locke, or even Hume.[21]

In chemistry, the same recurrence of interest in form as appeared everywhere. Modern chemistry, from Boyle and Hooke to Lavoisier, was a science of mathematical proportions, giving rise to the table of elements seen today in every elementary science classroom. Contemporary chemistry focuses on the more detailed differences in the *forms* or shapes of molecules. We learn that differences between left-handed and right-handed forms of the same element or compound can have important effects on its behavior, even though in the proportions of matter, the two forms are identical.

Such phenomena are nowhere more important than in cellular biology. The shapes and forms of organisms do not depend on mere matter. Rather, the critical phenomena arise from the shapes or forms of DNA, acting as a template for amino acids and proteins, and at the cellular level, the shapes or forms of cells and cell assemblages, which lead to differences in the activity of identical strings of DNA under the influence of the precise cellular environment.[22]

These effects are especially critical in the central nervous system. In the human brain, as in mammals more generally, neurons are inhibited or excited as chemicals known as receptors are matched to corresponding molecular shapes. The effect of a given neurotransmitter can be blocked or imitated by other molecules of similar shape. It is not the matter but the form that counts. Indeed, measurements of neurotransmitters and other active chemicals, on the assumption that they should reveal linear relations between quantities and responses, have again and again been confounded by the complexity of a system that does not seem to be essentially linear in nature.

In analyzing linear systems and universal relationships, modern science focused on what were, in a sense, the easy cases. In the linear domains of a complex system, results are indeed highly predictable. The difficulty is that one does not know the limits of predictability and therewith the limits of linearity. For the ancients, because scientific understanding did not exclude the zone of chaos in Figure 3.1, it was most important to seek patterns (Figures 3.3 and 3.4) instead of linear sequences and predictions (Figure 3.2). For such a science, prediction and quantitative measurement are less interesting that qualitative description and evaluation. If this account is correct, we must reverse the traditional understanding: instead of assuming that the ancients explored relatively simple problems that were a special case of the broader framework made possible by modern mathematics and natural science, it would follow that the moderns explored a relatively limited case that can be found within the broader framework defined by ancient science.[23]

The One and the Many: The Place of Values in Ancient Science

For ancients in the Socratic tradition, as in Euclidean geometry, it is possible to distinguish accurately among "natural kinds," and therefore to know—in principle if not always in practice—the nature of a thing. If so, a properly trained observer can also know what is "good" for each kind of thing. Scientific knowledge (fact) could therefore be the basis for moral judgment (value).

Whatever the differences among the ancients, *all* of the major Greek thinkers, from Thales and Pythagoras to Socrates and his followers, taught that their quest for knowledge illuminated the way an intelligent person should live. For Sophists like Gorgias, Antiphon, and Thrasymachus—as for Epicurus and his school—the end is individual pleasure, albeit variously defined; for Plato and Aristotle, as later for Cicero, Seneca, and the tradition we call classic, the end is moral or intellectual virtue. The ancient scientists thus demonstrate, by their deeds, that it is possible to have a science that engages in rational discussion of the proper ends or values of human life.

It modernity, radical doubt has become pervasive as the premise of all scientific methodology. In the tradition of Descartes, Hume, and Kant, we can never know the essence of things but merely formulate contingent hypotheses about their appearance. For the moderns, it follows that the properties used to describe reality are merely human conveniences or customs, albeit useful ones insofar as they can predict or construct outcomes. Values—our judgments of right and wrong—thus appear to be different from scientifically demonstrable facts: if so, the ancient mode of reasoning is based on the so-called naturalistic fallacy, confusing statements about what "is" with what "ought to be."[24]

To understand the ancient view, we must first realize that it was based on a different logical problem. Whereas moderns have been concerned by the logical relationship of *Is* and *Ought* or the so-called fact–value dichotomy, the ancients worried about the relationship of the "one" and the "many." How do we know when two different things or experiences are really the same? Not only is an adequate understanding of the relation of the one and the many fundamental to all forms of science, but it provides a more satisfactory answer to the dilemmas of moral choice than the presumed logical gulf between facts and values.

The question concerns the diversity of the phenomena that present themselves to human thought and experience. Perhaps the best-known ancient statement of the problem is the formulation of Heraclitus: "Every day the sun is new."[25] The things we see are constantly changing, constantly in flux. If one can't step into the same river twice, then all visible things are different; the impression of unity is merely imposed on this flux by the human mind. Or, in Protagoras' famous formulation, "Man is the measure of all things."

The difficulty is real: science cannot be based on naïve sense impressions, for we all know of appearances that turn out to be deceiving. Even some of the ancients were able to discover that, contrary to appearances, the earth is spherical and orbits around the sun.[26] For Heraclitus, as for many other ancient Greek thinkers, the critical question for all scientific or philosophic reason is the identification of that unity without which the multiplicity of perceptions and events is mere confusion. Having challenged experience as the ground of knowing, Heraclitus concluded that the unity or truth hidden in the multiplicity of appearance could be found in *logos*: rationally discoverable patterns, of which the most fundamental was probably the discovery that "all things are an equal exchange for fire and fire for all things"—that is, the relationship between mass and energy underlying physics since the discoveries of Einstein, Heisenberg, and Bohr.[27]

The logical issue of how to distinguish between one and many in appearances as well as in reasoning was thus central to ancient scientific thought. It is, in particular, the basic philosophic issue according to the Platonic dialogues.[28] According to the Platonic Socrates,

I suggest that the way to reflect about the nature of anything is as follows: first, to decide whether the object in respect of which we desire to have scientific knowledge, and to be able to impart it to others, is simple or complex; secondly, if it is simple, to inquire what natural capability it has of acting upon another thing, and through what means; or by what other thing, and through what means, it can be acted upon; or if it is complex, to enumerate its parts and observe in respect of each what we observe in the case of the simple object, to wit what its natural capacity, active or passive, consists in.[29]

The first step to scientific knowledge is to distinguish the "simple" from the "complex," or the one from the many; only then can a naturalistic science of effects or functions be possible.

This emphasis in Plato's work can be seen elsewhere. It is the basis of the principle of contradiction, on which all logic rests. In a famous section of the *Republic*, the issue is put as follows:

Socrates: Is it possible that the same thing at the same time and with respect to the same part should stand still and move?
Glaucon: Not at all.[30]

Any science, whether of nature generally or of human nature more specifically, rests on this principle; in the context of the passage just cited, for example, it serves as the foundation for Socrates's distinction of three parts of the soul and therewith the discovery of the principles on which justice rests.

Consideration of the problem of the one and the many—or the difference between simple and complex things—goes to the heart of the relationship between ancient and modern science. It would appear that moderns often make assumptions about the simplicity or uniformity of phenomena and theoretical concepts when the nature of these phenomena and concepts is open to question. On three distinct levels, it can be wondered whether modern science has imputed unity to that which is complex. Let us consider briefly the logic of the one and the many as it relates to time, to things, and to the city.

1. *Time.* For modern science, each moment in time is essentially the same as every other moment. This basic insight makes possible the mathematical analysis of continuous movement, opening an enormous range of subjects to scientific exploration and technological exploitation. At the limit, this assumption breaks down. One cannot equate our common experiences of time with what mathematical physicists call a "singularity" or the first instants after the "big bang" in prevailing cosmological theories (which explain why the world is composed of matter rather than anti-matter).[31] This difficulty arises in many other domains as well.[32] The many instants do not *always* make one type or kind of temporal event—and even if the exceptions are few in number, some of these exceptions are of overarching importance.

2. *Things.* The modern view also leads to the assumption that a uniform category of phenomena is identified by precisely defining a set of identical features or consequences. For the ancients, it is always a question whether the many instances of a category are really one. In recent years, scientists have often discovered that apparently identical things reveal vital differences on closer inspection: atoms of different isotopes, redundant codons in DNA, functionally identical polypeptides or proteins, types of neurons, individuals of a species, and so on. At one extreme, it is misleading to assume that all categories are merely conventional; at the other, it is incorrect to accept superficial appearances as adequate for a characterizing of homogeneity. For instance, animal species are indeed natural kinds, but each species contains immense genetic and phenotypical variability.

3. *Communities.* The oneness of the city was, for ancient political thinkers, as much of a problem as the unity of natural phenomena was for the cosmologists. The "noble lie" in Plato's *Republic* indicates the inevitability of this problem, for citizens do not grow from the earth speaking diverse languages and adopting diverse customs. Although Socrates's proposals seem to be a means of resolving the problem, his device of merging the city and the family is not practicable and indeed is ultimately comic.[33] The existence of communities is by nature, even though the precise form and substance of the city's laws and character are to some degree conventional.

At all three levels, the difficulty with the modern understanding is not that it is logically "wrong" but merely incomplete. This incompleteness helps to explain the problem of *Is* and *Ought*. In the realm of fact (*Is*), the identity of a moment in time, of a thing, or of a community, is always to some degree uncertain. In principle, we can know only probabilities. Natural kinds are therefore best understood as distributions or collections of similar things whose similarity is a matter of more and less; as Aristotle puts it, "However much all things may be 'so and not so,' still there is a more and a less in the nature of things."[34] Translated into modern mathematical terms, the normally distributed curve is a better description of a category than any supposed definition of its "essence."[35]

If facts generally concern a statistical distribution (or, in Aristotle's terms, that which is "for the most part"), values concern injunctive obligations or duties in individual situations. Each of us is desperately concerned with his or her own life (and death), not the average statistics for the population. Each individual has duties to specific members of a family and society, and these circumstances are not deducible from "average" or typical cultural norms that apply more or less across populations. What we call questions of ethics and morality are often prudential or practical choices in specific cases, and such cases often come to our attention because they reflect two or more conflicting principles that have become widely accepted in our culture.

Moral obligations or values thus concern individual or specific cases, whereas scientific descriptions or facts need to be understood in terms of populations and probabilities. But from a population of events, one cannot predict any single outcome with absolute certainty. As the joke goes, no human family has 2.4 children. Hume's skepticism needs merely to be reformulated in terms of the logical problem of the one and the many, which was at the heart of ancient thought, to become intelligible and consistent with contemporary scientific findings. It follows that moral obligations or values cannot be *logically deduced* from factual propositions but that factual or scientific propositions can and must inform the judgments about moral obligations.[36]

Stated at this level of generality, a resolution of the dilemmas of fact and value is not too helpful (even if intellectually satisfying). To go further, it is necessary to focus on specific problems. Human obligations differ if they arise in the family, in cultural or social relationships, or in political societies. Naturalism can help to identify not only the character of human relationships in each of these contexts, but also the necessary contradictions among them.[37] Science, whether ancient or modern, cannot go further, for in the individual case of each human life, decisions are needed—

and these decisions are precisely what determines the extent of human excellence.

For the moderns, nature can—in principle—be "controlled" or conquered. For the ancients, this was impossible. Contemporary ecologists are increasingly aware of the dangers of the modern view. Consider the example of antibiotics and disease. From a linear point of view, medical science discovers that a specific antibiotic can "cure" a given viral disease; if so, one can calculate the ultimate conquest of the disease as more and more humans use the antibiotic. There is only one problem. While humans are using the antibiotic, the virus need not be passive. The virus, like any living species, can evolve resistance to the antibiotic. Indeed we should expect it to do so.[38]

The consequence is simple. The perspective of ancient science, which—contrary to popular beliefs—had no difficulty in predicting evolution and change,[39] would not assume the linear relationship between a medication and the cure for the disease. Since nature can be understood but not controlled, the ancients expected the natural things to be in continual flux as they relate to human medicine. Hence one would expect the inevitability of a new form of pathogen and indeed of new plagues; the more humans seek to conquer such possibilities, the more inevitable they become because the very action of humans leads to increased variability of the pathogens.

The "good times"—the epochs when the human arts control nature—exist, but they are limited in duration. Humans do not progress; rather, history follows patterns that may approximate bifurcations, Lorenz attractors, or even oscillatory cycles. Just as civilizations rise and fall, so improvements in public health cannot be maintained indefinitely.[40] The domain that we think is controlled by modern science might thus be merely a small area in a broader picture, in which nature controls us—not the reverse. To understand natural phenomena and the place of humans in the world, therefore, modern science needs to be complemented by, if not subordinated to, the methods and assumptions of ancient science. Locke was perhaps too hasty in dismissing the concept, which he attributed correctly to the "old philosophers," that virtue is "the highest perfection of human nature."[41]

PART TWO

THE NATURE OF FACTS AND VALUES

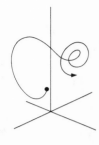

4. Is Knowledge Possible?
Subjectivity and the
Rejection of Science

For many contemporaries, human thinking is always subjective and relative. Are all ideas and judgments merely a reflection of the experience, self-interest, or customs of particular individuals living in a specific time and place? Do value preferences influence the choice of what is accepted as a fact?

Relativists have asserted that knowledge is merely "any collectively accepted system of beliefs." [1] For the extreme skeptic, all thought is subjective because humans have no means of transcending the parochialism and cultural values of their individual life experiences. Although such an attack on all scientific or religious claims to knowledge goes back to the earliest epoch of Greek philosophy, the so-called postmodern charge that subjectivity renders truth inaccessible has become widespread in the philosophy of science, epistemology, and cultural anthropology [2] as well as in the humanities more generally. [3] Without addressing this challenge directly, one can hardly propose a return to the perspective of ancient science, according to which the discovery of objective knowledge can provide a factual basis informing our understanding of what is good for humans "according to nature."

My answer to relativism will be divided into three parts: first, the reasons for rejecting the argument that humans cannot know anything because experience is entirely subjective (chap. 4); second, the demonstration that humans can communicate by means that are not entirely shaped by language and culture (chap. 5); and finally, the evidence that scientific theories are not merely reflections of social prejudice and political ideology (chap. 6).

What Is True? Three Models of Knowledge

Debates concerning human knowledge and the foundations of morality are often bewildering.[4] To simplify the discussion, I distinguish three principal approaches: intuition, empirical hypothesis testing (or verification), and pattern matching.

Intuition. According to this category of thought, all experience is subjective. The measure of when a human can be said to know anything is hardly more than the feeling or intuition of understanding. This perspective has been taken by scholars who see all scientific research as inherently limited by the cultural or political presuppositions of the scientific community, by humanists who doubt the possibility of knowledge (especially about human affairs), and by those for whom religious doctrines and scientific theories have the same status.

Among literary critics, writers, and artists, the reliance on intuition or feeling is perhaps not surprising. But similar views have been presented in the social and natural sciences as well as in philosophy. The German sociologist Max Weber, for example, stressed *verstehen*—intuitive understanding—as the ultimate criterion of knowledge of another culture; more recently, Clifford Geertz has championed a similar view by focusing anthropological research on "thick description" of cultural practices.[5] In one interpretation, the primacy of subjectivism is a corollary of Bohr's quantum physics and the discovery that the observation of a particle modifies its position or velocity.[6]

Hypothesis Testing, or Empirical Verification. The second approach, more widespread in the scientific community itself, has been formulated in a number of different ways.[7] Accepting Hume's skeptical conclusion that scientific theories can claim only provisional truth-value (see chap. 2), supporters of this view argue that scientific hypotheses or laws of nature are only valid insofar as they accurately predict or deduce observed events. In the version of this philosophy of science developed by logical positivists, scientific knowledge of this sort is possible with regard to natural things and can be extended to the study of human society.[8]

In a somewhat different view, widely associated with Sir Karl Popper, the truth of propositions rests on the extent to which they have not yet been falsified by empirical observation or experiment.[9] Based on a stringent interpretation of Hume's skepticism, Popper proposed that we can claim to know only those things that *can be*, but *have not yet been*, shown to be contrary to the evidence.[10]

Finally, many scholars use a more pragmatic approach, describing both general, or "covering," laws and falsification as principles used by the scientific community in its quest for verifying the truth of statements about the world. For example, Richard Lewontin, Steven Rose, and Leon Kamin offer "an essentially operational definition that is appropriate for assessing statements of truth in science, at least. In this definition, a true statement about an event, phenomenon, or process in the real material world must be (a) capable of independent verification by different observers; (b) internally self-consistent; (c) consistent with other statements about related events, phenomena, or processes; and (d) capable of generating verifiable predictions, or hypotheses, about what will happen to the event, phenomenon, or process if it is operated upon in certain ways—if we act upon it." [11]

Whatever the label used, supporters of these approaches to knowledge usually assume that scientific hypotheses (statements in the form "If x, then y") can be used to generate predictions or tests that will show whether a theory is true or not. Knowledge, then, is associated with reproducible theories, propositions or predictions that are formulated (often mathematically) in a way that can be tested by others using procedures stated in advance. Of particular importance is the reproducibility of results: individuals working independently need to find evidence consistent with predictions that could conceivably be wrong. In a sense, therefore, knowledge is created by generating precise propositions that can be falsified or verified. It is not hard to see the connection between these approaches to human knowledge and the modern fact–value dichotomy to which I will return in chapter 7.

Pattern Matching. A third and somewhat different perspective can be described as a pattern-matching view of thought. This approach to knowledge differs from intuition by asserting that the match between a pattern of thought and "reality" can give rise to scientific knowledge; it differs from verification because it denies that a formal law or hypothesis functions as an algorithm to predict outcomes that are then falsified.

In this mode of knowing, a pattern is distinguished because it resembles a prior expectation or form. We recognize a right triangle and a square because of their shapes, not because we can write down and verify the rules for constructing them. Often, there may be no single defining characteristic or operative rule that distinguishes one category from another, yet we have little difficulty in knowing the difference between them.

The recognition of animal species provides an excellent example of pattern matching. Although there is no single definition of what a species is, biologists can list a set of criteria for deciding whether two animals are of differing species; the more of these criteria that are met, the more likely it is that we are observing actually different species.[12] In general, therefore, it is easy to recognize a cat or a dog, even though there are likely to be difficult cases at the margins of each species. Anyone who enjoys watching birds is familiar with the process by which multiple cues of shape, color, voice, bodily movement, and habitat can be used to identify an ambiguous sighting.[13]

Although the view of knowledge as pattern matching can be traced to Plato, it is associated with recent work in history of science, neuroscience, and artificial intelligence. Neuroscientists find that, for such cognitive tasks as recognizing a face or remembering a scene, the human brain functions as a complex pattern-matching system. At a more theoretical level, a pattern-matching approach to knowledge also answers issues posed by cosmologists and physicists who have puzzled about the congruence between human thought and the structures of the world that are neither visible nor easily verified.[14] And as I will show, this approach helps to overcome the otherwise intractable debates between critics and supporters of very existence of science and objectively valid knowledge.

Intuition and the Critique of Science

Postmodernism has been associated with a widespread challenge to scientific objectivity based on intuition. Over the past seventy-five years, many scholars have explicitly challenged all claims to objectively valid knowledge about reality. If religion, literature, art, science, and philosophy are merely socially constructed and culturally determined, no one perspective can be taken as simply true.[15] Even more important, science itself appears as a mere reflection of the economic and ideological needs of bourgeois society.[16]

This challenge is often based on the explicit assertion that no human thought can transcend the status of an intuition or belief.[17] The position is perhaps clearest in the movement in literary criticism known as deconstructionism, as exemplified by the late Paul de Man. In this view, it is impossible to have knowledge about literary texts or indeed about any other human utterance. Because we can never know what the author or speaker intends, meaning is always constructed anew by the reader or listener. This act involves a destruction of the appearance of fixed or lasting meanings; hence the name deconstruction.

Such perspectives have challenged the traditional curricula of our universities. The so-called traditional canon of literature or philosophy is viewed as inherently biased because it excludes perspectives from other cultures—and from the disenfranchised within Western civilization. Since all knowledge is relative, no one perspective should dominate our university curricula. The criterion for accepting an idea or an interpretation becomes an intuition or a feeling of what is, at best, utility and, at worst, ideology and political correctness.

For example, the contemporary philosopher Richard Rorty has suggested that subjective bias can be minimized by expanding one's horizon to include a diversity of cultural and individual experiences. Rorty's practical means to this end is telling: one's vocabulary should be enlarged as much as possible. For this purpose, Rorty encourages philosophers to read novels and critical commentaries. Science does not seem to have a privileged claim to truth. On the contrary, since creative fiction often includes a variety of words concerning human affairs, it can be preferable to narrowly scientific or technical prose, whose language does not teach us directly about the diversity of human experience.[18]

This approach seems to deny the difference between science and other modes of thought, including dreams and works of art as well as theological doctrine and religious belief.[19] Reflection indicates, however, that contemporaries like Rorty have adopted the philosophic premises of Locke and Hume, according to which all thought is based on experience. If so, only a greater breadth of experience can claim to improve a person's understanding of the human condition. At one level, no scientist would disagree: if what is natural is what is most probable or common, as Hume suggests, then it is essential to have a broad sample of events before coming to conclusions about human nature. But at another level, Rorty's view implies that its own foundations are as weak as those of any other approach.

As I will show in chapter 7, not every criticism of science as contingent and subjective is derived from the idea that human thought can never go beyond intuition. For example, according to some Marxists and sociologists of knowledge, contemporary natural science uses concepts and methods that reflect capitalist or bourgeois society; their argument rests on a claim to have knowledge about human history. In contrast, the extremes of intuitive postmodernism reject the possibility of any objective scientific knowledge. Why has this latter view become so widespread?

History may have contributed to the attack on science in the name of intuition. For many, belief in science and technology has been undermined by the horror of Hitler's "final solution" and the fear of nuclear annihilation. The Nazi program of systematically eliminating the Jewish population of Europe amounted to the industrialization of murder, using modern technology to make genocide more efficient. Truman's decision to use the atomic bomb on Hiroshima and Nagasaki revealed the danger that the products of science could destroy life everywhere, removing the illusion of security and progress on which modern society has rested. In both cases, science and technology seem to have served ends so contrary to expectations that the entire modern view of the world was called into question.

For many intellectuals, the Holocaust was particularly difficult to understand. How could one of the most "advanced" scientific nations, whose universities had been among the most prestigious in the world, engage in such barbarism? Even more disturbing was use of supposedly scientific theories as the ideological justification for constructing factories for killing humans, known euphemistically as concentration camps. The Nazis used seemingly scientific theories of evolution and eugenics to define some groups as not truly human, and hence suitable for elimination by genocide. It is hardly surprising that scientific theories in general—and biology specifically— were accused of causing or encouraging the ultimate in human evil.

A generation later, the younger generation found its future under a mushroom-shaped cloud. At the height of the cold war, modern physics and its technological exploitation threatened to produce nuclear annihilation rather than the "conquest of nature" for the "relief of man's estate" promised by Bacon. These fears were exacerbated when American leaders used the calculus of deterrence, supposedly based on a science of human behavior, to justify the war in Vietnam.

If the pretentions to science are responsible for such evils and dangers, surely it is preferable to rely on less dangerous approaches to human thought. At the root of the contemporary challenge to scientific knowledge is the intuition that modern science has been responsible for untold harm. Such a profoundly skeptical assessment of modern science and technology, increased by fears of environmental pollution and by economic uncertainty, is by no means foolish.

It does not follow that knowledge is inaccessible merely because science has been systematically misused. On the contrary, such an argument against science presupposes some form of knowledge or truth about the effects of human behavior. For example, the charge that twentieth-century theories in biology were somehow responsible for the genocidal policies of the Nazis is itself a causal claim. As Lewontin, Rose, and Kamin put it, "the sorry history of this century of insistence on the iron nature of biological determination of criminality and degeneracy, leading to the growth of the eugenics movement, sterilization laws, and the race science of Nazi Germany has frequently been told."[20] Indeed, it illustrates nicely the problem of causal inference emphasized by Hume. The Nazis used scientific or pseudoscientific theories and advanced technologies to implement a policy of killing over six million people. Doesn't it follow that scientific thought, by "leading to" these horrors, is somehow responsible?

The causal relationship between science and the evil consequences attributed to its use is subject to Hume's skepticism. Before accepting intuitively obvious explanations, we need to consider the facts. In the twentieth century, two major totalitarian regimes engaged in industrialized murder and technological control over society: Stalinist Russia and Hitler's Germany. Under Stalin, Marxist–Leninist ideology accepted only theories of environmental determinism: Lysenko's biology was used to criticize genetic research, and theories of innate or biological causes of human behavior were rejected in favor of explanations based on social class and economic power. Under Hitler, Nazi ideology developed doctrines of racial purity and genetic determinism, providing the mirror image of Soviet ideology. Both regimes committed systematic atrocities of unparalleled violence. Both were anti-Semitic. It is hard to conclude that the excesses of twentieth-century totalitarianism are entirely due to *either* biological *or* economic theories.

History teaches that humans have often engaged in racism without relying on science at all. In the United States, such unscientific (if not antiscientific) xenophobia is symbolized by the Know-Nothing movement: long before Darwinian biology, strangers were attacked because they were convenient political targets. Much the same can be said of the explosion of ethnic conflict in Eastern Europe following the fall of Soviet communism. The reasons for organized hostility to outsiders are doubtless associated with socioeconomic and cultural tensions, to be studied by historians and social scientists. The biological theories used—or rather, *misused*—by Nazi ideologists to justify the Holocaust were probably not the sole, nor even the primary, cause of genocide.

At the same time, the intuitive reaction contains an element of truth. The rationale used to legitimate a policy may make it more or less difficult to criticize; humans seek to construct explanations or interpretations of their own behavior which then can justify it. Scientific ideologies may therefore be criticized because they make it easier for elites to do evil or dangerous things. The problem with intuition is that, to go beyond vague fears, one needs something like scientific knowledge.[21]

The defect of the intuitive view of racism and xenophobia can be seen from a recent experiment. According to the charge that biological theories are responsible for social bias, the rejection of outsiders should be associated with ideas about strangers. Yet a recent study shows that the process can be just the reverse. A sample of American adults saw television excerpts of unidentified leaders from Germany, France, and the United States;

FIGURE 4.1

*Cognitive Judgments, Emotional Responses, and Thermometer Ratings
of American, French, and German Leaders*

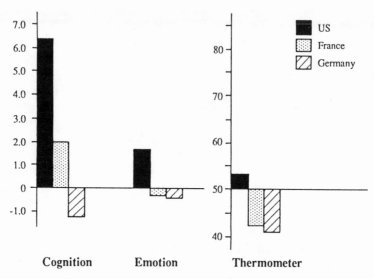

$N = 84$. Cognitive ratings: average summation of 11 bipolar Semantic Differential Scales. Emotional responses: "net warmth" (average of positive emotions minus negative emotions as verbally reported on 6 unipolar self-reporting scales. Thermometer: overall rating on 0–100 scale. Source: Warnecke, 1991 (see n. 22, this chapter), Tables 6.13, 6.15, 7.1.

**Variation across three nationalities of leader significant at $p < .001$ (Cognition: $F = 54.2$; Emotion: $F = 18.2$; Thermometer: $F = 26.4$). For all three measures, one factor ANOVA shows that the average rating of American leaders was significantly different from ratings of French and German leaders ($p < .05$), whereas mean ratings of French and German leaders were not significantly different.

Note: the reader is invited to compare the relative ease of understanding the differences between responses to American leaders (black bars) and those from France (gray) or Germany (diagonal lines) with his/her own reactions to Table 4.1. The way data is presented by scientists can be used to illustrate the three modes of knowledge distinguished above.

when unfamiliar officials were seen without the sound, so that neither nationality nor identity was known by the viewer, the emotional responses and judgments elicited by the Germans and French were more negative than reactions to Americans. In short, without knowledge of who was a fellow-citizen, viewers reacted with unconscious prejudice against the outsiders (see Figure 4.1, in which the black bars represent responses to American leaders in terms of cognitive traits, viewer's own feelings, and an overall 0–100 thermometer scale that usually predicts voting choices). Yet when the same viewers saw these television images *with* the sound, so that nationality was self-evident, the effect disappeared.[22]

Something in the nonverbal behavior or body language of foreigners seems to elicit negative reactions. When the same viewers were aware that someone was foreign because they saw excerpts with sound, the hostility did not occur. According to the common opinion, reflected in the attack on biological theories as a cause of Nazi racism, conscious beliefs and conscious intentions lead to racism and xenophobia. Such intuitive explanations cannot explain why unconscious hostility to foreigners disappeared as soon as the viewers knew the leader's nationality. The experiment reveals both the *importance* of the prevailing beliefs (which can either discourage racism and xenophobia, as in the contemporary United States, or encourage it), and the *limits* of their causal efficacy (the roots of racism may be deeper than the arguments used to justify it).

Other researchers have confirmed the principle underlying this example. Although much of our understanding of the world does not flow from conscious decision, it is the methods of science that demonstrate the accuracy and limits of such preconscious or intuitive modes of knowledge.[23] The intuitive challenge to science is trapped in a vicious circle: intuitions are indeed sometimes true, but to understand when and why this is so, it is necessary to rely in other modes of knowing.[24] Today we need to explore the claims of scientific knowledge more fully.

Modern Science, Verification, and Knowledge as Making

The practices and beliefs of the modern scientific community rest on a characteristic approach to knowledge. Natural science is usually said to discover laws of nature. These general propositions (whether viewed as empirically demonstrated knowledge or as tentative hypotheses) are human constructs that can be tested by experience. The experiment just mentioned tests the hypothesis that hostility to outsiders has emotional or nonverbal antecedents that could be either encouraged or discouraged by conscious beliefs and social practices.

The scientist thus tries to take a specific case—whether the motions of the planet Jupiter, the functions of the neurotransmitter serotonin, or the use of racist ideas by the Nazis—and explain its origins or effects by a more general rule. These propositions lead to predictions. *If* the theory is true, *then* we should see a specific result. The outcome of the test is not known in advance. In the study mentioned above, the American viewers might have responded in a similar way to all leaders regardless of their nationality—or even preferred French and/or Germans to their fellow citizens.[25] Because the method is open to contrary findings, if the predicted result is not observed, it is assumed that the theory needs modification. Experience can thus change or refine statements about the world.

The scientific propositions that are tested have a specific character. According to one interpretation, they are rules or algorithms, predicting outcomes in a class of events under specified conditions. The discovery of such scientific hypotheses or general laws is seen as a human creation or construction. The scientist must formulate the rule according to which the event could have been produced. Specifying the conditions for empirical verification requires a factual test, in which these conditions are either observed or experimentally generated. In either case, the scientist tries to formulate the rules or regularities that produced observed phenomena.

In this sense, the modern perspective considers knowledge as a form of making. As one philosopher of science puts it, "Scientists want theories which explain and predict the world, which allow them to manipulate the world in a variety of ways. Theories or paradigms merely compatible with experience are of no use whatsoever." [26] Humans can discover the rules by which nature is governed only by an active process, not by passive contemplation; they need to manipulate the world. Presumably this is why successful scientists are said to be creative.

This constructive view of laws of nature may help to explain the role of mathematical formalization in modern science. As noted in chapter 2, Leonardo had already said that "no human investigation can be called true science without passing through mathematical tests." [27] Why should this be so? For most contemporaries, the mathematics of probability has become the response to Hume's skepticism. If a prediction could occur by chance, one has little reason to believe that it reflects an understanding of the causes at work. When watching someone flip a coin, I call heads before the coin falls; if the prediction is confirmed, can I claim that I *caused* the outcome?

Results that could have happened by accident do not give us confidence that our explanation is correct. The prediction of a single coin toss has a fifty-fifty chance of being right. Outcomes that could only have occurred in one case out of a hundred seem to give more confidence in the hypothesis or rule. By increasing the frequency of observations, the scientist makes it easier for the testing process to contradict the prediction. In the experimental results displayed in Figure 4.1, the overall pattern of responses to leaders of the three nationalities would have occurred by chance only once in a thousand times. Even so, of course, the result may not be true. But repeated confirmation by different experiments, particularly if conducted by different scientists, increases one's confidence that the rule has been accurately stated. [28]

The measure of confidence that a finding did not occur by chance is

TABLE 4.1
*Significance of Pair-wise Differences
in Responses to Leaders from US, France, and Germany*
(Image-Only Excerpts, Low Status Leaders)

	US–French		US–German		French–German	
	Fisher PLSD	Scheffe F-Test	Fisher PLSD	Scheffe F-Test	Fisher PLSD	Scheffe F-Test
Cognitive judgment						
Female (p = .0001, F = 23.4)	3.0**	7.6**	3.0**	23.2**	3.0**	4.1**
Male (p = .0001, F = 33.1)	2.4**	10.9**	2.4**	32.8**	2.4**	5.9**
Emotional self-report						
Positive emotion						
Female (p = .0032, F = 5.8)	.7**	3.8*	.7**	4.8**	.7	.1
Male (p = .0002; F = 8.5)	.7*	3.3*	.7**	8.3**	.7	1.1
Negative emotion						
Female (p = .0041, F = 5.5)	.8**	4.8**	.8**	3.4*	.8	.1
Male (p = .0336; F = 3.4)	.6*	2.7	.6*	2.4	.6	0
Thermometer rating						
Female (p = .0001; F = 10.6)	.4**	8.4**	.4**	7.5**	.4	0
Male (p = .0001, F = 16.1)	.3**	7.2**	.3**	15.4**	.3	1.5

One-factor ANOVA.
**Difference between two countries is significant at $p < .01$;
*Significant, $p < .05$.
Note: Although this table presents information about the same results as those presented in Figure 4.1, it is almost impossible for the nonspecialist to understand. As one reader commented, "These tables are very hard for me to understand. Without interpretation, they do more harm than good."

called a test of significance, a term reflecting the substitution of Hume's skepticism for the naive confidence in intuition. The presentation of results in scientific discourse therefore takes a more statistical form than is common in ordinary writing. Moreover, the statistical significance can be computed for each detailed comparison that makes up a broader pattern. Since many nonscientists have found mathematics unpleasant since the fifth grade, this reliance on numbers may have reinforced the division between scientists and humanists, but this hardly constitutes a good reason to deny the objectivity of scientific findings. For example, most readers of this book will probably find the graphs in Figure 4.1 easier to understand than the statistics in Table 4.1, even though, for a specialist, it is the latter that make the experimental results significant.

The logic of using mathematical tests to assess the validity of knowledge, while derived from the thought of Hobbes, Descartes, and Hume, is most widely associated with Newtonian or classical physics. Long the reigning paradigm of science, the theories of classical physics assumed that natural phenomena have objectively defined properties (mass, velocity, position) that are not dependent on the observer's position or actions. All objects measured as having the same mass, velocity, and position should behave in the same way, no matter which scientist studies them. It is only on this assumption that general laws of nature can make reasonable predictions.

The objects that have been measured do not, however, always have the properties we expect: the coin being flipped may have two heads. As Hume noted, we can never prove that a relationship observed in the past (or in one group of phenomena) will apply in the future (or in other similar cases). For example, in studies of human behavior, it is assumed that the subjects of an experiment or public opinion poll resemble the general population of interest. Even with an attempt to control for what is called sampling error, however, unsuspected factors often make an experiment unrepresentative of what it is supposed to test. For instance, in the experiment summarized in Figure 4.1, the viewers' personality was measured, and, unexpectedly, it was found that the means of recruiting subjects had excluded some kinds of people.[29]

It is also important that different observers find the same thing. The case of cold fusion described in chapter 2 illustrates the point. Some philosophers speak of scientific knowledge as intersubjectively verifiable: one person often cannot see or understand another's intuitions, but anyone can be convinced by evidence based on repeatedly observed confirmation of a prediction. To be sure, training in scientific methods may be needed. But people are, in principle, equal in their ability to seek this training. The verificationist view does not treat truth as the preserve of a specialized sect or class of wise men, priests, or shamans.[30]

The assumption that things are homogeneous and measurable makes possible equilibrium models, specifying the stable state of a system under given conditions. Such models, common in such diverse fields as classical physics and contemporary economics, often reveal unsuspected relationships in complex systems. Moreover, in both the physical and social sciences, this method of analysis facilitates technological innovations, using scientific knowledge as a means of transforming the conditions of human life. When knowing takes the form of constructing general rules or laws and making conditions to verify them, the process seemingly leads to increasing power over nature.

These features of modern scientific knowledge have, however, been viewed as a reflection of cultural norms rather than an image of the truth. As some philosophers of science emphasize, the "objectification" of things and the assumption of their measurability lie at the root of commercial, capitalist society.[31] Understanding human knowing and action as a kind of making is the foundation of industrial productivity and modern technology, not a universal assumption of all human societies. The social dimension to science also can be seen as a product of a particular culture. The methods of falsification or verification that can be practiced by any trained scientist democratize science, reflecting the equality of rights and social status in Western societies since the fall of the ancien régime.[32]

Although postmodernist critics argue that this poses a deep problem, virtually all of the assumptions at issue have been challenged by contemporary natural scientists themselves. I will return to the claim that the findings of science are vitiated by cultural or ideological bias (chap. 6). For the moment, it should suffice to note that scientific theories of relativity, quantum mechanics, and chaos theory have undermined the notion of universal laws of nature, pointed to the possibility that form is prior to matter, and therewith suggested the continued relevance of ancient science.

Ancient Science, Pattern Matching, and Knowledge as Seeing

Philosophic debates concerning the nature of truth might be even more difficult to resolve if the modern process of verification were the only means to scientific knowledge. The existence of a different approach to science clarifies many problems. Although represented by Plato, Aristotle, and the tradition of ancient science and philosophy, pattern matching is, as we shall see, by no means limited to the past. What, then, is the difference between the modern scientific process of hypothesis falsification or verification and this older view of knowledge?

In the ancients' approach, knowledge is typically the discovery of a pattern. The ancient Greeks did not use the number zero and had not discovered the calculus. Their model of mathematical rigor was a proof in Euclidean geometry.[33] The essence of this procedure is the reductio ad absurdum.

If a circle is a line equidistant at all points, and if a line is drawn bisecting the circle through the center, the area of the two segments is equal (Fig. 4.2a). This *must* be so because any difference between the two segments would prove either that the line a–b does not go through the center or that the line representing the circle is not equidistant at all points from the center (Fig. 4.2b). Like modern science, the demonstration is reproducible. Unlike modern science, it does not depend on the empirical measurement of the circle. The figure drawn can be very approximate (the rough sketch of Fig. 4.2b) and yet demonstrate the point as well as a more carefully drawn image (Fig. 4.2a). Images only *represent* the form of a circle and merely help us to *see* the relationship being examined.

These characteristics of ancient science are clearly set forth in the works of Plato. For example, when describing the methods to be used in analyzing

FIGURE 4.2
The Geometrical Model of Knowledge

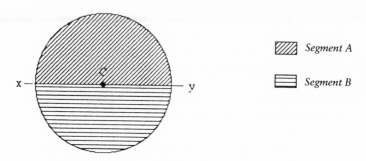

Segment A

Segment B

a: A carefully drawn image of the proposition that, given a circle with its center at point C, the diameter *x–y* divides the circle into two equal segments (*A* and *B*).

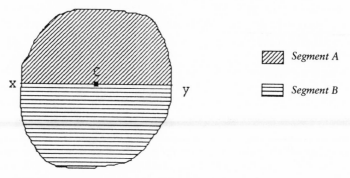

Segment A

Segment B

b: A roughly drawn image of the same proposition.

human nature, Plato seeks a method to reach "a more precise agreement" when appearances are contradictory and confusing:

> if someone were to say . . . that tops as wholes stand still and move at the same time when the peg is fixed in the same place and they spin, or that anything else going around in a circle on the same spot does this too, we wouldn't accept it because it's not with respect to the same part of themselves that such things are at that time both at rest and in motion. But we'd say that they have in them both a straight and a circumference; and with respect to the straight they stand still since they don't lean in any direction—while with respect to the circumference they move in a circle; and when the straight inclines to the right, the left, forward, or backward at the same time that it's spinning, then in no way does it stand still. (*Republic*, 4.436c–e; ed. Bloom, p. 115)

The ambiguities of intuition can be overcome by using the principle of contradiction. "It is impossible that something that is the same, at the same time, with respect to the same part and in relation to the same thing, could ever suffer, be, or do opposites" (ibid).

Much has been said about Plato's theory of knowledge and his concept of Ideas or Forms. Some of these descriptions are simply wrong: Plato was not an idealist for whom the evidence was irrelevant.[34] On the contrary, the immediate sequel to the passage just cited resembles the premises of modern verification: "Let's assume that this is so and go ahead, agreed that if it should ever appear otherwise, all our conclusions based on it will be undone" (*Republic*, 437a; p. 115). Although Plato and the ancients more generally were aware of the importance of empirical evidence, they do differ from moderns in an important way: the scientists and philosophers of antiquity emphasized knowledge of the form and nature or essence of a thing rather than the rules for making or predicting visible phenomena. Knowledge is the recognition of patterns, not the creation of scientific laws.

This difference in method helps to explain how ancient science could study that which is good for a thing, whereas modern scientists focus on laws predicting motion or change. The difference is manifest not only in mathematics (Greek geometry versus modern algebra) but in areas as distinct as physics (the ancients sought to define the basic elements; moderns formulated laws of motion that could predict the orbit of a planet or the acceleration of a falling body), biology (the ancients emphasized the reasons that animals have certain shapes or parts; moderns considered theories of evolution and development to be more important), and political science (the ancients defined the kinds of regimes and above all the "best" regime; moderns looked for individual "rights" and other universal "principles" of political life).

Thinkers in antiquity differed enormously on how they interpreted the world: The Sophists rejected justice as a natural standard for human behavior, whereas the Socratics usually asserted it; the Epicureans rejected political life as a diversion from individual happiness, whereas the Stoics insisted on the primacy of virtue and obligation; Platonists asserted the reality of the forms, whereas Aristotelians questioned it. But with only marginal exceptions, the ancients used methods to discern knowledge and truth different from those of modern scientists.[35]

Pattern matching is not merely a quaint historical feature of ancient Greek and Roman thought. Recent studies of the human brain have shown that perception is typically a process in which sensory input is compared to remembered features of the form or pattern of a thing. Particularly in gestalt perception, holistic features are seen all at once rather than produced by a rule or algorithm akin to a computer program. Many other features of human thought resemble linear processing, including human speech. Hence the difference between knowledge as pattern matching and as a rule or algorithm is found within the brain itself.

This duality has sometimes been traced to differences between the two sides of the brain, with the right cerebral hemisphere supposedly specialized in holistic perception and the left in linear processing (whether in motor coordination or in language).[36] Even within the cortex, specific cell types (such as the magno neurons in the visual cortex) and structural areas (such as the inferior temporal lobes) have been identified as particularly devoted to the recognition of form.

The modes of knowing distinguished above correspond not only to different methods of scientific discovery but to distinct processes in the brain. Intuition or "feelings" correspond to the emotional responses in the limbic system, now demonstrated as essential for learning and memory. Verification or falsification based on hypothesis testing reflects the type of linguistic and mathematical information processing characteristic of the left hemisphere and, in sensory processing, what has come to be known as connectionism. Pattern matching and holistic assessment of form characterizes the functions of the right hemisphere and, more broadly, the global integration of the brain's parallel processing of multiple cues in the environment.

The practical importance of considering how the brain works is illustrated by our daily experience of recognizing another person. We see a picture and say, "I know who that is." How do we do this? Neuroscientists have found that facial recognition results from a distinct pattern of neuronal activity. A face isn't remembered by a single cell but by an ensemble of neurons firing in a characteristic pattern.[37] In other areas as well, it seems that distributed *patterns* of activity in multiple centers of the brain are associated with learning and memory.[38] I will return to the contributions of neuroscience in chapters 6 and 8, but it should already be clear that human thought is a compound of three approaches that have traditionally been viewed as mutually exclusive.[39]

The findings of contemporary science can thus contribute to a solution of the problem of infinite regress which has long taxed theories of knowledge. Relativists have repeatedly asserted that all science is vitiated by the influence of individual experience and social conditions. But differing conceptions of knowledge seem to reflect processes that occur simultaneously in the brain of every thinking human. If so, the conditions in which something is learned (the context of discovery) will typically be different from the truth of knowledge (the context of justification). Relativists are correct in noting that particular circumstances allow humans to discover features of experience and reality, but it does not follow that intuition is the only criterion of knowledge and truth. If anything, reanalysis of the history of science in the light of contemporary studies of cognition suggests that, of the three modes of knowing, it is pattern matching which may be the most general or fundamental.[40]

Intuition, the Socratic Question, and the Dialectic

In this chapter, I have focused on three methods of understanding knowledge: intuition, hypothesis testing, and pattern matching. Insofar as each rests on a major functional system of the human brain, all three seem to have a valuable function—but this function is understandable only if the role of each of the three is given its due place. Humanists or poets, modern scientists, and ancient philosophers all sought the deepest understanding of human life and nature. Each, however, has tended to use a different standard of knowledge or truth (see Fig. 4.3).[41] Those who assert that all knowledge is subjective have posed an important problem that needs to be resolved before we can hope to understand the potentiality and limits of science. But intuition itself cannot be the answer.

The postmodern critique of both tradition and science, which seems to lead to relativism and nihilism, is but the latest example of a persistent theme in Western intellectual history. Repeatedly, an intuitive sense of truth and fairness has been used to challenge claims to knowledge based on religious faith or scientific reason. Although intuition itself cannot resolve the question of human knowledge, the critique of science or custom in the name of subjective meaning can have a useful function.

My defense of both verification and pattern matching should therefore not obscure the positive role of intuitive challenges to supposedly objective knowledge. Skeptical intuition lies at the heart of Socrates' endless questioning. It is no accident that Socrates was condemned as an unbeliever and that the main school traced to his tradition was called the Academic Skeptics. We misunderstand the character of the origins of Western political thought if Plato's presentation of Socrates, particularly in late dialogues like the *Republic*, is confused with the historical Socrates; other ancient writers, from Xenophon to Aristotle, provide us with plentiful evidence that the historical Socrates did not accept Plato's theory of the Forms or Ideas, whereas all sources agree that he questioned everyone.

The historical Socrates, as we know from ancient sources, had studied the natural sciences of his day. He turned away from these studies, not only because they seemed useless in dealing with human problems but for the deeper reason that he could not be sure of gaining *knowledge* from such a process.[42] Indeed, Aristotle is quite explicit in claiming that this Socratic turn actually harmed some branches of the scientific study of nature.[43] Yet

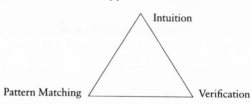

FIGURE 4.3
Three Approaches to Truth

	Intuition	Verification	Pattern Matching
Act of knowing	Feeling	Making	Seeing
Nature of things	Heterogeneous	Homogeneous	Heterogeneous
Measurability of things	No	Yes	Sometimes
Reproducibility of knowledge	No	Yes	Yes
Human abilities to know	Unequal	Equal	Unequal
Dominant cerebral process	Feeling	Linear processing	Gestalt processing
Dominant neuro-anatomical structure	Limbic system	Left hemisphere	Right hemisphere

in overall terms, it is hard to deny that what we think of as the Western thought was first institutionalized by the tradition inaugurated by Socrates and continued by Plato, Xenophon, Aristotle, their schools, and their followers.[44]

As a mode of knowledge, intuition is not limited to its role in supporting skepticism and relativist criticisms of scientific objectivity. Although intuition has also supported religious beliefs, it plays a central role *within* the processes of both ancient and modern science. If the feeling of dissent or doubt lacks legitimacy, knowledge becomes mere ideology. It is only against continued questioning as a legitimate activity that the discovery processes of either verification or pattern matching can claim to attain to knowledge.

Postmodernism, like earlier forms of skepticism, can therefore have positive functions. On the one hand, it is but the latest version of the challenge to unquestioning belief, countering the tendency for opinions to come to serve the self-interest of those in power; on the other, it has a particular role to play in an epoch of rapid change. Insofar as theoretical ideas and concepts are shaped by social realities, an epoch of rapid change might make possible new insights. The approach of deconstruction is not entirely inappropriate as a way of challenging opinions and beliefs that reflect social structures in the process of disappearing.

The problem with postmodernism, then, is not its existence; rather, it is the assertion that intuition alone matters and that, on the basis of intuition, all claims to knowledge are equally fallible and equally vitiated by human pride and self-interest. As a challenge, relativism can be like the Socratic question, forcing us to adapt our answers to changing times and new discoveries. As an answer, relativism is extraordinarily dangerous, for it breeds precisely the insecurities and fears that sociologists find at the heart of the turn to authoritarian regimes.

According to a tradition going back to Socrates and his students but as current as any good classroom, knowledge rests on dialogue. Knowledge becomes suspect when it becomes authoritative doctrine, taught on the assumption that only one perspective can possibly be given credence. The Platonic dialogues give us an image of this dialectical view of knowing. It may well be that pattern matching—reasoned assent based on a comparison between experiences, intelligible forms, and explanations—is the most general way of conceptualizing knowledge. But intuition and empirical verification cannot be excluded from the dialogue without putting one individual in a position to suppress the statements of anyone who disagrees.

Two questions need to be asked if science is to adjudicate between competing modes of knowing. First, is human communication possible without relying on conventional definitions and criteria that already establish the basic premises of discourse (including the categories of right and wrong, good and bad, that we call values)? And second, even if it is possible to communicate knowledge through science, do the methods and standards of science serve the purposes of social power and self-interest rather than those of objective or disinterested truth? Chapters 5 and 6 address these fundamental questions.

5. The Nature of Social Communication: Nonverbal Behavior, Cognition, and Innate Ideas

In order for science—or any other form of dialogue—to be meaningful, the participants need to share an understanding of the world. Doctrinaires and absolutists of all sorts seem to feel that there is only one "correct" way to think. Relativists assert that meaning is imposed on a meaningless universe by human will, stipulation, or custom. What is the nature of social communication? Is all thought socially conditioned onto minds that are "blank slates" or do humans have "innate ideas" which provide a natural foundation for common understandings?

For centuries, philosophers have speculated about these questions. Today, however, the question is being answered by concrete scientific evidence. Experimental findings indicate that the brain has inborn potentialities that are shaped by learning. In other words, the answer to "nature versus nurture" is *both* nature (innate ideas) *and* nurture (individual experience and social custom).

When discussing the foundations of meaning, philosophers have typically used logic to analyze what humans can know about material things based on examples from physics, chemistry, or other natural sciences.[1] Now, in contrast, discoveries in ethology and cognitive neuroscience make it possible to reassess the objectivity of our knowledge of the world, and the best evidence concerning the nature of meaning may come from perceptions and judgments of human beings, not of physical objects.[2] Two changes are therefore involved: first, concrete evidence can replace abstract speculation; second, ideas about other people are more important than those about inanimate things.

Nonverbal Behavior, Emotion, and Social Cognition

There is good reason to focus on social communication as the foundation of human knowledge. Social relationships between individuals are characteristic of many animal species; human society has natural origins that are revealed in the behavior of primates and other mammals. In these species, social signals have evolved as a means of conveying information to other members of the group.[3] Among humans, facial expressions are displays of emotion that have evolved from similar nonverbal cues in closely related animals; they too communicate information to others.[4] Although cultural norms, personality, gender, individual experience, and social environment shape perceptions and judgments of nonverbal signals, they are as natural to humans as traits like bipedal walking or copulation.

The system of nonverbal communication common to all humans contradicts the argument of skeptics in the tradition of Locke and Hume, for whom differences in human custom and belief prove that there are no innate ideas. For example, a relaxed, smiling face is identified as an expression of happiness by people in every known culture, elicits positive responses in newborns, and is even exhibited by blind children who cannot have learned it by watching others. Among nonhuman primates as well as humans, such facial cues are both an expression of happiness and a reliable social signal of reassurance. Similar information is communicated by facial displays of anger and threat or of fear and evasiveness. Nonverbal behavior communicates an innate idea of dispositions to help, to harm, or to defer to others—the building blocks of all social interaction.

Nurture matters, too. Both expression and interpretation of nonverbal behavior are shaped by culture and individual experience. Humans—and even some monkeys and apes—can learn to use these cues deceptively: one can smile to lead another to think one is feeling happy when that is not the case.[5] Such behavior shows, however, that the basic idea transmitted and elicited by the signal is naturally shared and communicable. Deception works only when the actor can reliably predict how the nonverbal cues will be perceived and decoded; the difference between honest and deceptive communications is that, in the latter, the expected response of the other becomes in itself part of the stimulus for the actor.

Many philosophic puzzles of epistemology can therefore be clarified by studying how living organisms "know" their worlds (a field that is sometimes called evolutionary epistemology).[6] In other animals as well as in humans, behavior is typically the result of the interaction of inherited components, environment, and individual experience. Innate ideas can thus provide the natural basis for communicating meaning among humans even though—or rather precisely because—they are shaped by culture and education.

Why then do individuals—and groups—differ so widely in response to the same social events? The question is difficult because the prevailing theories in modern social science do not account for the way differences of culture, personality, and gender cut across conventional categories like social class, education, and income. Although individual experiences are the cause of differing perception and behavior, cultures vary by shaping roles and expected responses. And while socioeconomic variables contribute further variation, there are some characteristics of personality and behavior that tend to differ between men and women. Nonverbal communication illustrates all of these complexities.

Human social interaction has its roots in dominance, reassurance, and competitive behaviors found in nonhuman primates, not to mention other social mammals.[7] Although many social patterns or group structures have been observed among nonhuman primates,[8] the adaptation of primate groups to environmental constraints[9] does not seem to have required extensive specialization in each species' behavioral repertoire. On the contrary, the nonverbal displays associated with reassurance, confidence, threat, attack, fear, or submissiveness are strikingly similar among nonhuman primates and across all human cultures.[10]

Although several types of nonverbal behavior—including vocal cues, body position, and body movement—are important social signals among primates, facial displays are of particular importance in humans and will be the focus of this discussion. Such facial cues are essential to the process of mother–infant bonding.[11] During early development, the smile is expressed under similar circumstances even by blind infants, confirming that it is not entirely reducible to cultural convention.[12] If a mother abruptly replaces smiling behavior with a neutral facial display when interacting with her four-month-old infant, the child finds the change aversive and, after smiling to elicit maternal reassurance, will cry with anger and frustration.[13]

As children mature, more of the facial display repertoire is used, and its role in social interaction becomes more complex. In peer groups of three- or four-year-olds, many of the social patterns found in nonhuman primates are expressed by human children.[14] Conflict over a toy or a space is at first likely to elicit agonic behaviors (anger, threat, or aggression; fear, evasion, or submissiveness), and this behavior is predictive of the outcome.[15] In older children, however, dispute settlement is more likely to entail reassurance, turn taking, and appeasement behavior.[16]

While one member of a group tends to become its leader, dominance among five- or six-year-old children is typically associated with the ability to use reassurance displays in dispute settlement and as a means of encouraging playful interaction. The most aggressive individual is usually not the leader (in general, among primates as in face-to-face human groups, the second-ranking male is more likely to engage in aggression than is the leader). The dominant child is the individual on whom attention is usually focused rather than the one insistent on an exclusive control of resources; among young children, girls are as likely to fill this role as are boys.[17]

Among adults, facial displays and other nonverbal cues continue to play a central role in expressing individual moods and organizing social behavior. Facial expressions of reassurance elicit positive emotion, whereas fear displays arouse fear; it is more difficult to condition a positive response to a fear display than to a smile.[18] The sight of another individual's displays of reassurance, threat, or fear induces the psychophysiological changes associated with each of these emotions.[19] And if an actor is verbally instructed to move one facial muscle at a time until a characteristic display has been unconsciously formed, these physiological attributes of emotional experience also result.[20]

The facial displays associated with basic human emotions and social cues are easily and reliably decoded across virtually all known human cultures.[21] In a series of experiments discussed in more detail below, we have found that expressions of happiness/reassurance, anger/threat, and fear/evasion can be defined on the basis of objective cue contrasts. Studies in France and the United States show that videotapes of known leaders exhibiting the three types of facial displays are reliably distinguished even if strong political attitudes influence the viewer's descriptive ratings. And in both countries, the three types of display elicit predictable patterns of emotional response, indicating that a leader's nonverbal expressions are indeed meaningful to citizens.[22]

Dominant adults characteristically exhibit responses that are associated with self-assurance and the ability to reassure others. In general, expressions of fear or submissiveness are not observed in successful leaders among humans or nonhuman primates; human observers usually respond with negative emotion when leaders show these fearful expressions.[23] Reassurance displays, in contrast, are particularly important elements of leadership behavior, particularly after arousing situations and social conflict; for humans, such displays generally increase positive emotion and reduce negative feelings.[24] The significance of anger/threat displays depends more precisely on context and appropriateness. If a human leader shows threat in response to challenge from an outsider, followers will be reassured; the same leader's display of unprovoked anger and threat toward others inside the group can elicit fear and may entail a loss of status.[25]

As might be expected in an evolved system of social signals, the primate brain has evolved structures that are specialized in the recognition of faces as well as in the decoding and response to facial display cues. Individuals who cannot name familiar faces (prosopagnosics) exhibit appropriate emotional responses on seeing a known face, thereby confirming that the memory of others need not be mediated by verbal information.[26] Similarly, differences in the emotional expressions of leaders could be accurately decoded by a patient who otherwise had difficulty with verbal memory.[27]

Localized structures in the brain are specialized in these functions of facial recognition and display decoding. Of special importance are the inferior temporal lobes, which mediate a pathway between the visual cortex and the limbic system that is basic for associative learning and memory. The face of another individual is recognized by the firing of a small number of associated neurons (ensemble coding), whereas specific cue contrasts like raising the head apparently elicit differential responses from single ("dedicated") neurons in the same area of the brain.[28] Individuals suffering damage to the temporal lobes often show deficits in social behavior: if the temporal lobes have been removed, even though other components of behavior are not affected, nonhuman primates are unable to form social bonds, and a human patient reported an inability to establish emotional responses to others.[29]

Identifying a known face, whether President Reagan (Figure 5.1a or b) or Clinton (Figure 5.1c or d), is thus a process of pattern matching. At the same time that we identify *who* we are seeing, however, we are also responding to that person's nonverbal behavior, for instance assessing a facial display as happy and reassuring (Figure 5.1a or c), fearful and evasive (Figure 5.1d), or angry and threatening (Figure 5.1b). The intuitive response to the sight of another thus requires both multiple pattern matching and conscious verification or comparison of expectations to what is seen. In short, the process of seeing another's face elicits all three knowledge processes discussed above in chapter 4.

It is therefore reasonable to assume that the nonverbal behavior of leaders might be a relevant cue in the formation of public attitudes toward rivals for power. Western literature provides ample evidence that leaders have traditionally attended to such behavior as a crucial element in establishing and maintaining status and power (e.g., Shakespeare, *3 Henry VI*, act 3, scene 3, lines 168–95; *Henry V*, act 4, scene 1, lines 103–11; Milton, *Paradise Lost*, book 2, lines 302–9). In the study of rhetoric, training in nonverbal behavior—especially in the appropriate facial displays of emotion—was once considered a necessary element in social and political success, and the importance of such nonverbal cues has been recognized by

FIGURE 5.1
Presidential Facial Displays

President Reagan

a: Happy/reassuring display

b: Anger/threat display

President Clinton

c: Happy/reassuring display

d: Fear/evasion display

famous philosophers[30] as well as in recent theories of human social and political behavior.[31]

These findings demonstrate that humans, like chimpanzees and other nonhuman primates, are naturally social animals for whom leadership is associated with a combination of reassuring and competitive behaviors. The Hobbesian or Lockean view of humans as atomistic and selfish individuals whose learning is the product of personal experience is inconsistent with ethological observation and evolutionary theory. The related Hobbesian assumption that all social cooperation needs to be deduced from the costs and benefits to selfish participants is equally open to question.[32]

Cognitive Neuroscience and the Modular Brain

Contemporary research on the human brain points to a more comprehensive approach to the way the brain integrates diverse verbal and nonverbal cues in meaningful social behavior. All areas of cognition exhibit the general characteristics found in our responses to nonverbal behavior. The resulting picture is one of complex interactions between innate and acquired components. Education and experience shape or modify a natural repertoire of perception and social behavior (instead of writing individually distinct messages on a blank slate, as the Lockean tradition would lead one to believe).[33]

While human behavior cannot be predicted solely on the basis of the way our brains process information, prevailing theories often make assumptions about human psychology that now appear to be false or misleading. Just as the Freudian approach to dreaming has been challenged by neurophysiological and neurochemical research,[34] the behaviorist model of the brain has been demolished by recent research. Unfortunately, the obsolete psychological models underlying political science, economics, and sociology have not been revised in the light of discoveries about the structure of the brain and the way it processes social information.

New research on the central nervous system establishes three main principles on which scientific models of human nature must be based: (1) the "modular brain" as a parallel processing system, (2) the essential role of emotion in learning and memory, and (3) individual differences in neuronal structures as well as in cognitive processing. These three principles characterize most features of human social behavior and thereby provide a basis for an evolutionary psychology that is scientific without being reductionist. Applied to the concrete phenomena of leader–follower relationships, they explain otherwise puzzling complexities in the perception and response to facial displays and other nonverbal cues. A word on each of these principles of cognitive neuroscience will therefore be necessary before showing how they explain individual and cultural or ethnic differences in social behavior and thought.

Perception, decoding, emotional response, memory, and verbal communication are based on highly specialized neuronal assemblies organized in a system of interrelated processing units, or modules. The brain is composed of localized and functionally specific structures that process different features of the environment (like the face or voice of another person) in parallel. Although the physical development and the response properties of these neuronal structures depend on the individual's experience, their activation is coordinated and integrated by the global organization of the modular brain. Hence it no longer makes sense to think of human thought and feeling as the product of an undifferentiated and totally "plastic" black box.

Each functional system uses specific neurons that are organized hierarchically to perform information-processing algorithms, many of which seem to resemble artificial intelligence models.[35] In the visual cortex, for example, stimulus properties associated with line, shape, color, motion, and spatial location are processed by specialized arrays of neurons. A specific object in the environment thus elicits parallel processing by a multiplicity of feature detectors, starting with highly abstract cues (straight versus curved lines, hue or intensity of color, direction of motion in the visual field, etc.) that are integrated in hierarchical feed-forward circuits, themselves strengthened or shaped by prior learning and experience. Hence, if a human subject is instructed to attend to either the shape, color, or motion of a stimulus, different sites in the visual cortex are activated.[36]

Holistic patterns are recognized and associated with each other by higher-order responses that activate arrays of cells or, in some cases, highly specialized neurons ("grandmother cells") that code for a complex pattern. If specific features of sensory input (e.g., curved lines) are artificially removed from the organism's environment during critical periods of development, the neuronal assemblies or modules that normally process this feature will atrophy or cease to be linked to the rest of the system. Conversely, experience and use strengthen and restructure neuronal pathways.[37] While the brain has innate response capacities, neuroanatomic structures are to some degree plastic and reorganized by experience. Interactional systems have thus replaced the older view in which perception and learning could be reduced to simple causal models.

Similar organizing principles are involved in processing language. Speaking, reading, and writing involve the integration of highly localized pathways of information processing.[38] Now described as "connectionism," the resulting model of the brain's cognitive functioning rests on the discovery that different stages of linguistic information processing (phonemic contrast, morphemic recognition or recall, prosody, grammar and syntax, etc.) are served by distinct processing units linked to form a hierarchical system. This view of the brain is not only consistent with models developed by some artificial intelligence specialists but has been confirmed by new techniques of recording cerebral activity such as MRI or PET scans and electrical recordings of neuronal activity (event-related potentials). When words are read silently with attention to their sounds, for example, areas of the auditory cortex are activated that do not respond in silent reading attending only to meaning.[39]

Equally important is evidence from studies of the developmental and acquired learning disabilities generally called dyslexia. Reading deficits that were once lumped in global categories like "word blindness" are now seen as heterogeneous, depending on the specific linguistic processing modules that have suffered damage in each individual. This new explanation, described as the disconnection hypothesis,[40] attributes deficits in reading and writing to neuronal damage to pathways linking localized linguistic processing structures or to one or more of these structures themselves.[41] For example, if the auditory cortex is disconnected from the linguistic memory centers, the resulting errors (writing "litsen" for "listen") are consistently different from those occurring when the deficit is in the linkage with the visual cortex ("lissen" for "listen").

Of what importance are these details? Because the "black box" model of the brain implied that attitudes and feelings are holistic phenomena, variations in human behavior were assumed to reflect "either-or" differences. Does language shape perception, or does perception shape language?[42] Is emotion mediated by cognition, or can feelings be immediate reactions independent of cognitive processing?[43] In political science, similar controversies have revolved around such questions as the determinants of voter choice: is it party identification, retrospective evaluation of those in office, assessment of the personality and character of rival leaders, or emotional reaction to the candidates that explains how citizens behave in elections?[44]

Such conventional ways of thinking about human cognition are contradicted by the theory of the modular brain. Parallel processing models of information processing extend to symbolic or cognitive sets; one explanation need not exclude alternative hypotheses. Indeed, research with "split brain" patients shows that a single individual's attitudes toward social objects can differ when responses are elicited from the left hemisphere as contrasted to the right; in one patient, President Nixon received a positive rating by one hemisphere and a negative one from the other half of the brain.[45] It is often impossible to define "the" feeling or attitude of a person or to determine whether one response is prior to another.

Multiple factors necessarily impinge on any single perception, feeling, or judgment. Different individuals (or one individual in different settings) will have varied ways of integrating emotion and cognition, attitudes and habits, lasting self-identification and situation judgments. Hypotheses or theories based on the observers' definition of holistic categories are contradicted by the way the central nervous system works, and methods of research need to take these findings into account.

For example, public opinion polls rest on the assumption that most people can describe accurately the factors influencing their judgments and behavior. The parallel processing model of the modular brain challenges this belief. Gazzaniga's research with split-brain patients reveals what he has called an interpreter module, distinct from linguistic areas in the left hemisphere. This part of the brain assesses the diverse responses of other information-processing modules, attempting to produce consistency among them. What psychologists have long discussed as responses to "cognitive dissonance" can now be studied directly with neuroscientific techniques.

Experiments with split-brain patients demonstrate how humans generate and express post hoc explanations of thoughts and behavior that reflect subconscious or linguistically inaccessible information.[46] Directly asking voters why they thought or acted as they did may not uncover the actual causal processes, particularly if the research has not been shaped by knowledge of parallel processing in the human brain. For example, although experimental subjects often assert that their cognitive judgments of politicians have not been influenced by their feelings, emotional responses are typically more important than conscious judgments in determining overall attitudes.[47] Little wonder the public opinion polls are so often unreliable.

A methodological point follows. Many social scientists measure the correlation between a single explanatory variable and a behavior response; for reasons discussed in chapter 4, if the correlation coefficient is significant, it is assumed that there is an effect worthy of report. Given the concept of the modular brain, such an overall statistical measurement sometimes obscures more than it reveals; methods need to focus on how diverse factors are integrated in varied conditions.[48] Not only is it imperative to consider how a multiplicity of independent variables relate to each other and to observed responses, but there is no reason to assume that these patterns are constant for different individuals or situations.

Emotion, Associative Learning, and Memory

The distinction between emotion and cognition is one of the overly simplistic dichotomies of past psychological theory that has been contradicted by cognitive neuroscience. According to some scholars, feelings are shaped by judgments and cognition; for others, emotion need not be influenced by consciously held attitudes. Research on the brain reveals that both processes occur in parallel, since there are at least two major neuroanatomical pathways linking sensory input with the limbic system (the center controlling the physiological responses we associate with emotion).

As Mishkin and Appenzeller demonstrated, one pathway (ventral) transmits impulses from the eye or ear to the sensory cortex, then to the temporal lobes and thence to the amygdala; another pathway (dorsal) processes sensory input through the neocortex before transmitting neuronal impulses to the hippocampus.[49] The first of these pathways seems likely to be less cognitively mediated than the second because it has fewer synaptic links between primary sensory areas of the brain and neocortex prior to reaching the limbic system.

For smell and perhaps for other sensory modalities, moreover, there is an even more direct pathway from the primary receptors (such as the olfactory bulb) to the amygdala and hippocampus that entirely bypasses the cortex.[50] Such structural features of the brain presumably explain the difference between first impressions and subsequent judgments, not to mention the immediate and vivid emotional responses to odor and the difficulty of describing in words such powerful and socially important sense impressions as those epitomized by Proust's account of the madeleine.[51]

The concept of a dichotomy between emotion and cognition is thus challenged by recent neuroscientific discoveries. Without processing by the amygdala or hippocampus, associative learning does not occur; memory also involves the activation of the emotional centers of the brain.[52] Now that it is possible to relate brain function to neuronal structure, categories of analysis need to correspond with the processes that actually occur in the central nervous system.

From the neuroscientific perspective, a principal distinction is presently recognized between declaratory memory (responses that a subject can make a focus of consciousness and explain verbally) and performance, or nondeclaratory, memory (automatic or habitual reactions of which one is often not fully aware). Declaratory memory can in turn be distinguished into factual memory (recall of specific information) and episodic memory (recall of specific events), each of which can be based on nonverbal mental images or on verbally stored and expressed information. Nondeclaratory memory includes motor coordinations (how to hit a golf ball or drive a car), rules of linguistic decoding and verbal performance, dispositional responses to specific stimuli, habitual perceptual routines, and the like. Each of these functional categories seems to correspond to localized neuroanatomical processes.[53]

What difference do these findings make for the study of human social behavior and leadership? There is no question that social and political attitudes are learned. But the way judgments are modified and activated depends, to a decisive extent, on the way the brain learns and remembers. For the last decade, social scientists have explored rational-choice models of behavior, cognitive theories, and sociological explanations that tend to ignore the role of emotions in human social behavior.[54] Since the emotions elicited by a given event vary from one person to another, such rationalist accounts of behavior tend to obscure the differences between people.[55]

Three specific consequences follow. First, the question can no longer be *whether* emotion is related to cognition, but rather *how* this relationship occurs. Second, it is not enough to focus on verbal or linguistic information; nonverbal sensory input, mental images, and symbols can be as important as verbal information and responses, especially insofar as many aspects of social behavior are not directly accessible to conscious verbal report. Finally, it becomes critical to distinguish carefully between the formation or change of attitudes in episodic responses to specific events and the lasting residue of these experiences (which can still affect attitudes and behavior even if no longer stored in episodic memory).

Individual Differences: Culture, Gender, and Personality

The concept of the brain as a modular or parallel-processing system in which emotions and cognition are closely linked challenges many conventional assumptions about human nature and social behavior. It leads, moreover, to a greater emphasis on individual variations in responses to what an observer describes as the same event. From a biological perspective, the structure and function of the central nervous system is a product of epigenesis—a process of individual development throughout life in which environment and experience shape neuroanatomical structures as well as hormonal or neurochemical systems regulating behavior.[56] What a social scientist assumes to be an identical stimulus or situation can represent different realities for people whose brain development or experience followed divergent epigenetic pathways.

Apart from the idiosyncratic events impinging on each individual throughout life, at least three levels of developmental differentiation are of primary relevance to human social and political behavior. Each can be viewed as a source of variation in the way individuals respond to similar social cues. With regard to responses to leaders, for example, each can be thought of as additional independent variables in a complex, multicausal system.

Culture.   From the perspective of the modular or connectionist model of the brain, cultural variation is associated with differences in the way the central nervous system processes and responds to environmental stimuli. Many of these features concern performance (nondeclaratory) memory, that is, learned responses that are not normally accessible to conscious recall. For example, after puberty the capacity to discriminate and produce speech sounds seems to disappear for languages that have never been learned or heard.[57] Rates and rhythms of speech also differ systematically from one language to another, and these rhythmic patterns seem to govern preferred rates of information processing in other sensory modalities; as a result, the presentation of visual images in TV newscasts in different countries seems to follow rhythms similar to those observed in the verbal speech of each country.[58]

Cultural anthropologists have, of course, stressed the verbal or conscious symbols that are the basis of the manifest rules and practices of a human society. But from an evolutionary point of view, these rules represent a narrowing of the range of individual responses, increasing the predictability of social interaction.[59] The shaping and reinforcement of selected neuronal assemblies and pathways, combined with the atrophy of unused connections or structures, provides a mechanism that explains sociocultural differences. In this way, it is possible to go beyond the sterile nature–nurture dichotomy to investigate complex interactions between the natural repertoire of social behavior and the specific patterns of each human culture.[60]

Nonverbal behavior seems to be particularly important as a component of the resulting differences between one culture and another. Preferred uses of social space, such as the typical distances between interacting individuals, vary across cultures, as do hand gestures and other patterns of body movement.[61] Similarly, the system of facial displays, especially those cues associated with leadership, dominance, and subordination, are shaped by cultural differences in expectation and behavior.

While our species has a natural repertoire of emotional and expressive displays, cultures differ in the way these communicative cues are decoded, even though we are often not fully conscious of the fact. In France, as was noted above, facial displays of anger/threat and happiness/reassurance by a leader elicit similar emotions, whereas viewers in the United States feel significantly less pleasure or reassurance when leaders show anger/threat than when they show happiness/reassurance.

Such cultural attributes as the egalitarianism and friendliness often attributed to Americans seem to reflect these differences in nonverbal social behavior. This explains why, when seeing silent images of unknown leaders from different countries, viewers spontaneously describe them by using the terms of common cultural stereotypes (see chap. 4). As a result, even when people have learned each other's language, cultural misunderstandings and stereotypes can arise from the misperception of nonverbal cues that have been differentially ritualized from one society to another.

Personality.   Culture is hardly the only source of human variability. No careful observer can fail to be struck by the immense individual differences in response to a single event, not only within a culture but even within each family. Siblings often differ remarkably; strangers, even from quite diverse cultures, often recognize more affinity with each other than each feels for close kin. How can this be?

Personality classification is notoriously difficult. In recent years, psychologists have proposed a number of classificatory schemes, not to mention the many evaluative systems used in counseling and therapy. Generally, however, models based on at least three major dimensions on which individuals vary have replaced the ancient classification of holistic personality types, according to which people were characterized as melancholic, choleric, sanguine, or phlegmatic.[62] In a modification of this newer approach, Cloninger has proposed a three-dimensional model based on neuroanatomy and neurochemistry that shows promise of integrating typologies of both personality and mental illness.[63] In this view, individuals vary on the dimensions of reward dependence (inner-directed or egocentric vs. outer-directed or socially dependent), harm avoidance (risk-taking vs. risk-aversive), and novelty seeking (passivity or activity in search of stimulation).

Individual differences in neurochemistry provide a plausible causal mechanism for such variations. Cloninger's model is based on differential activity in three main neurotransmitter systems, each of which can be thought of as "tuning" different neuronal structures or pathways and hence modulating specific perceptions or behaviors.[64] Cloninger attributes differences in the harm-avoidance dimension to the serotonergic system, variations in reward dependence to norepinephrine and its associated neurochemicals, and variation in the novelty-seeking dimension to the dopaminergic system.

The Cloninger model clearly simplifies more complex phenomena in the central nervous system. Each of the postulated neurotransmitter systems consists of a main neurotransmitter (dopamine, serotonin, or norepinephrine) that is dependent on numerous other chemicals (precursors, agonists, antagonists, reuptake agents, and other polypeptides that influence synaptic activity). Moreover, each system interacts with the others and each is influenced by additional neurochemical systems (such as the insulin regulation of glucose turnover, melatonin regulation of light response, estrogen regulation of the female menstrual cycle, etc.). Despite these complications, neurochemical imbalances and dysfunctions have now been implicated in a number of behavioral and personality traits, including suicide, alcoholism, seasonal affective disorder, and impulsive homicide or arson.[65]

Since the activity of many neurochemical systems is sensitive to experience, social situations, diet, and other environmental factors, behavioral responses dependent on the activity of dopaminergic, serotonergic, or norepinphrinergic pathways are usually labile rather than fixed. At the same time, however, genetic differences in baseline levels of various neurochemicals in each major system (precursors, agonists, antagonists, or other related peptides) can explain the heritability of personality traits that are modulated by neurotransmitter activity.

The otherwise puzzling evidence that personality traits depend on both inherited and experiential factors thus reflects an underlying system associated with neurochemistry. Such an explanation probably explains why personality types in dogs and monkeys seem to resemble those found in humans.[66] There is thus considerable evidence that variability in personality traits rests on natural mechanisms that humans share with other social mammals.[67]

For some variants of such conditions as schizophrenia and alcoholism, possible genetic markers associated with neurochemical disturbances have been identified—though in each case there seem to be multiple causes for observed behavioral syndromes, and each genetic marker seems associated with vulnerabilities rather than narrowly determined phenotypical traits. Similarly, behavior and social environment have been shown to influence the level and activity of neurotransmitters and hormones, such as increases in serotonin when an individual becomes socially dominant[68] or heightened cortisol levels under stress.[69] Individuals thus differ in their social predispositions, but individual experience usually modifies the responses of the underlying physiological systems.

However classified and whatever the mechanisms that produce them, human personality differences seem to be predictable and stable. Extensive population studies show that, of the potential sources of individual variation in such traits as sociability, cooperation, risk taking, and novelty seeking, little can be attributed to culture, whereas roughly similar pro-portions of variance are due to heritability and to individual life history.[70] While such statistical estimates have often been criticized on methodologi-cal grounds,[71] they are consistent with an epigenetic approach to human psychology and confirmed by experimental studies in mammalian behavior genetics.[72]

Individual differences in responding to nonverbal cues during social interactions seem to be a proximate mechanism underlying human per-sonality. At four months, infants differ in the latency and intensity of their responses to an abrupt change in the mother's facial display behavior; at seven months, these individual differences persist.[73] Children who show reticence or enthusiasm in initiating social interactions with an unknown peer at 2½ years exhibit similar traits at the age of 7; these differences, as measured by both gross behavior and physiological responses, are highly stable for those at the extremes of the population distribution but labile near the mean.[74]

In some extreme cases, personality disorders have been attributed pri-marily to dysfunction in the perception or response to nonverbal social cues. Autism may be associated with an inabiity to decode the facial dis-plays and nonverbal cues of others.[75] A recent study of the "failure to thrive" syndrome discovered that the affected children show a dispro-portionately high intensity and frequency of facial displays of fear and sadness.[76] Whether or not a neurochemical "tuning" model can account for these observations, it seems plausible to assume that differential re-sponsiveness to nonverbal cues could be one of the proximate mechanisms underlying personality differences. If so, we should expect that personality will be related to differences in social and political responsiveness but that this connection is mediated by differences in the perception of identical social cues.

Gender. A third major source of behavioral differentiation in social behavior concerns females and males. The controversies surrounding this factor are apparent in the debates concerning the appropriate word to describe it: for some, it is a question of "sex," whereas for others—par-ticularly if they claim that all such differences are shaped by culturally variable roles—it is important to speak only of "gender."[77] However ex-plained or labeled, few can now deny the overwhelming evidence that men and women persistently differ in their responses to similar situations, in their modes of social interaction, and in their attitudes to many political issues.[78]

There is, of course, an enormous literature on the character of these dif-

ferences as well as their possible evolutionary origins, ontogenetic development, and shaping by culture and experience. At the biological level, recent
research has shown how the basic physiology of males and females arises
from processes in fetal development that are under hormonal control.[79]
While usually ignored by social scientists, these findings have surprisingly
great relevance for human social behavior.[80]

Whatever the contributions of nature and nurture, males and females
process information and respond to others in somewhat different ways.
Properly conceptualized as two normal distributions with slightly different means, overlapping without coinciding, gender thus concerns probability differences rather than a holistic or typological distinction. Although
women and men have similar potentialities and responses for most variables, average rates of maturation, specific cognitive or physical abilities,
personality traits, and behavioral responses often differ on average by gender. As Beach put it a generation ago, most mammals are bisexual but
not equipotential in the performance of the behaviors usually attributed to
males and females.

Females are, for example, more likely to engage in affiliative behaviors,
whereas males have a somewhat greater tendency to initiate competitive
or aggressive interactions. This does not mean that females are passive or
nonaggressive but rather that their competitive behaviors are more likely
to arise as a consequence of preexisting social bonds, whereas male cooperation and competition are more opportunistic or situation-specific.[81]
While some scholars emphasize the extent to which these attributes are
shaped and reinforced by the norms of Western cultures, others point to
the origins of such tendencies in the behavior of other primates and even
other mammals.[82]

Of specific relevance here is the differential extent to which females integrate diverse elements in information processing and social behavior. Best
known in Carol Gilligan's analysis of the role of context and personal commitments in the moral judgments of women,[83] this factor can be traced in a
wide variety of cognitive and emotional responses. To cite but one example
of central importance to the role of innate cues in social communication,
males and females seem to respond somewhat differently to nonverbal displays of emotion.[84] Moreover, these differences sometimes contradict the
widely held sexual stereotype that women are more "sensitive" than men:
while females seem to be more responsive to the facial displays of infants,
one study showed that males were more accurate in recalling the facial
displays of political leaders.[85]

Gender differences have also been found in the way males and females
combine diverse social cues. In the modular brain model, one could say

that females integrate different functional processes more fully, whereas parallel processing or selective attention is more characteristic of males. At the physiological level, for example, there is evidence indicating greater functional integration of the hemispheres among females.[86] Mechanisms that might conceivably account for such sex-related differences include hormonal effects on neuroanatomy either during neonatal development or maturation, other neuroanatomic differences (such as the response capacities of the corpus callosum), differential levels and activity of neurotransmitter and hormonal systems, or the effects of differential experience in neuronal development and synaptic linkage.[87] Indeed, it is conceivable that many or all of these factors may exist to different degrees for some traits or in some individuals.

These gender differences have implications for the way social cues are processed. Verbal messages, nonverbal displays, images or symbols, prior attitudes and established emotions all influence the way an individual responds to a political speech, debate, news story, or TV advertisement. The hypothesized male–female differences in information processing imply that females should integrate diverse cognitive and emotional attributes more fully, whereas males should be more likely to respond independently to distinct cues. Experimental evidence confirms this prediction: males' responses to facial displays are more likely to depend on the display cues themselves; whereas, among females, emotional responses tend to interact with the viewer's prior attitude to the individual leader, the leader's status, and other contextual cues.[88]

Facial Displays and Political Leadership

We have an innate basis for interpreting the behavior of others that is shaped by individual experience and culture. Because emotions and learned associations interact when humans respond to our inherited repertoire of social cues, experimental research on individual differences in perception and behavior should focus on meaningful situations of real importance. Under contemporary circumstances, known political leaders are virtually the only people for whom the same behavior can be used to elicit comparable but diverse responses across a population.[89] These studies therefore illuminate the interaction of nature and nurture in human behavior generally, in addition to explaining leadership in contemporary societies.

When close-up images of today's leaders are shown on TV, their facial displays elicit emotions and influence attitudes in ways that can affect outcomes and shape contemporary institutions.[90] In a month-long cross-cultural sample, the face of a political leader was shown during 14 percent of the average newscasts on French TV, 17 percent in the United States, and 30 percent in West Germany.[91] A series of experiments, which I started in 1982 with colleagues and students at Dartmouth, has shown how these episodic events can affect the viewers' emotions, impressions of leaders, and potentially lasting attitudes.[92]

Our findings show that nonverbal display behavior is potentially of great importance in leader–follower relationships but that the effects are extremely complex. Viewers accurately describe different facial displays: average ratings of excerpts objectively defined as showing happiness/reassurance, anger/threat, fear/evasion, or neutral behavior were significantly different, with the scales congruent to the type of display rated higher than other descriptive categories. When rivals are shown during a single experiment, particularly at election time, descriptions are correlated with the viewer's established political attitudes; even then, however, objective differences in nonverbal behavior are clearly perceived by the average citizen.

Each type of display has a distinct pattern of effects on the viewers' emotional responses.[93] Broadly speaking, happiness/reassurance elicits positive feelings (but not negative ones), whereas fear/evasion is most likely to produce negative feelings in viewers. Patterns of response to anger/threat are intermediate: whereas there was no significant difference between the positive feelings during happiness/reassurance and anger/threat in our French study, Americans felt similarly when watching anger/threat and fear/evasion, each of which elicited less positive and more negative feeling than did happy/reassuring excerpts. In other words, viewers in both countries respond similarly to facial displays of happiness and fear, but an angry leader produces positive feelings in France and negative ones in the United States.

These episodic emotional responses while watching each excerpt depend very much on the viewers' prior attitudes. As a rule, viewers who support a leader are more influenced by the differences in a leader's display behavior, whereas critical viewers are less likely to respond very differently to display cues. Happy/reassurance displays have somewhat different effects for supporters (for whom these excerpts elicit strong positive emotion) and for critics (whose negative feelings are neutralized by these displays). Fear/evasion elicits negative feelings from supporters and critics alike, whereas the patterns of response to anger/threat were more complex, depending on attitude, nationality, and even media condition.

The emotions felt during exposure to TV scenes of leaders can be shown to have potentially lasting effects on viewers' attitudes. In one study, displays that were inserted in TV news stories had attitude effects for some viewers twenty-four hours after the last exposure, showing that the effects of nonverbal cues can last long enough to be further reinforced and become part of durable judgments and feelings. Particularly at the outset of a national election campaign, the mere exposure to excerpts of rivals—especially those who were little known—may significantly influence viewers' ratings of a leader and subsequent voting behavior.

A number of additional factors also enter into the system by which nonverbal cues communicate feelings and potentially modify attitudes. Among these variables are (1) channel of communication; (2) performance style of the leader; (3) intensity of display; (4) viewer's pretest attitude to the leader, party, ideology, and other opinions; (5) viewer's gender; (6) personality or sensitivity of the viewer; (7) cognitive information; (8) culture; (9) socioeconomic status and ethnic background of viewer; (10) status of leader; (11) competitive versus noncompetitive contexts; and (12) framing or cueing by TV announcers.[94]

These experiments show that innate cues in human social communication do not have identical effects for everyone or in all circumstances. Whereas modern social scientists have sought universal or invariant laws of behavior modeled on the laws of Newtonian physics, an approach to human thought and behavior based on the life sciences leads to a better understanding of individual and cultural variability. Such a naturalistic approach, characteristic of ancient science as articulated by Plato and Aristotle, thus overcomes limitations in the modern theories based on Locke's tabula rasa model of the brain.

Nature interacts with nurture, even though both modern natural scientists and postmodern skeptics have assumed that there is a gulf between them. If past experience is a guide, I predict that at least one reviewer of this book will criticize me for failing to disentangle nature and nurture. Such a mistake would prove either that the reviewer didn't understand anything in this chapter or didn't bother to read page 87. Scientific evidence often fails to convince scholars as well as ordinary readers.

In a time of rapidly accumulating scientific discoveries, resistance to new ways of thinking can be a critical problem. Like the rest of us, scientists often reject uncomfortable facts. For some critics, this is evidence that all scientific theories reflect the self-interest of the dominant groups in society. Is science itself something other than a disinterested search for the truth? This charge of relativism needs to be answered, for it is on a different plane from the views rejected in the preceding two chapters.

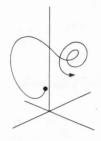

6. Is Science Relative?
Culture, History, and the
Structure of Scientific Theories

Evidence that shared meaning can be based on an innate system of social communication answers one objection of relativists but leaves another open. Cultures not only shape the meaning of such nonverbal cues as facial expressions; they also frequently adopt beliefs manifestly contrary to scientific knowledge. How do we know that the understandings of scientists are any less contingent than the beliefs of other cultures?

Research in the life sciences provides three good reasons to take this question seriously. First, evolutionary theory assumes that humans, like other animals, will typically seek their own adaptive advantage or self-interest; from a biological point of view, there is no reason to assume that scientists are more altruistic than other humans. Second, the neuro-scientific findings summarized in the preceding chapter reveal unconscious mechanisms that resolve inconsistencies in perception and other forms of cognitive dissonance; what Gazzaniga calls the interpreter module could easily account for biases in the scientists themselves as well as in the average citizen's opinions. Finally, the existence of nonverbal displays has a paradoxical effect, often making self-deception the most effective means of deceiving others; that scientists are persuaded of their own objectivity is hardly the last word.

Are scientific theories relative to the society or culture in which they emerge? In the sociology of knowledge it has been argued that theories about human nature and the world around us are reflections of the values and beliefs of those with power and influence. More specifically, it is often claimed that a naturalistic or biologically inspired theory of human behavior reflects unstated ideological commitments.

I cannot address this question without changing method. Theories from different historical epochs need to be compared to see whether their substantive content is consistently associated with political or social bias. This procedure has an added advantage, however, because it will also provide a useful perspective on the theories of human nature and society that are most directly associated with the ethical and political values.

The results show that similar scientific concepts have been elaborated in very different social and cultural contexts. The validity of a theory is independent of the circumstances of its discovery. And apparently contradictory scientific views can each contain an element of truth, pointing to the need for dialogue and openness to diverse perspectives.

How Is Science Related to Society?

Postmodernists claim not only that humans cannot *know* anything but also that all knowledge is relative to the society in which it is discovered and accepted. These two arguments are different—and to some extent contradict each other.[1] For example, one scholar attacking scientific objectivity has written: "There is much evidence that features of culture which usually count as nonscientific greatly influence both the creation and the evaluation of scientific theories and findings."[2] Obviously, such a statement presents itself as knowledge based on the facts.

Although some postmodernists themselves make this argument in a way that is logically self-contradictory, it does not follow that cultural relativism is false. As philosophers of science have long noted, the origin of a theoretical proposition is different from its truth. Examination of the charge that scientific theories and practices reflect social, political and cultural values is particularly necessary before it can be asserted that science can be a source of criteria for judging right and wrong. If the relativists are correct, such an endeavor would be circular.

It is impossible to deny that, like modern science, many features of ancient science reflect the cultural and political norms of society in which they emerged. Consider some of the obvious evidence for the relativist position.

- Ancient philosophers and scientists did not imagine that scientific knowledge was akin to making; in antiquity, productive labor was generally done by slaves, whereas education was primarily directed to the citizen. As a result, ancient science was divorced from technology, and science was typically viewed as the preserve of an elite.

*forms vs
algebra*

- Ancient philosophers and scientists assumed that the nature of visible things is heterogeneous; Greek and Roman societies were polytheistic. As a result, in antiquity the primary form of mathematical reasoning was the geometry of different forms rather than arithmetical calculation or the algebraic rules governing all motions and bodies.
- Ancient philosophers and scientists did not use the term "laws of nature" to describe abstract rules governing the visible world; the Greek city-state was small and parochial, with each city having its own patron gods. Even where empires existed, they lacked the extensive central bureaucracies of a modern state.

Similar comparisons between modern science and the societies in which it has flourished have been outlined in chapter 2. From Weber and Marx to Sartre and contemporary historians of science, it has been apparent that Western science was not likely to have emerged in another time or place.[3] History thus suggests that the practice of science always reflects to some extent social realities.[4]

If so, can science claim objectivity independent of time and place? Paradoxically, any answer implies that observers in one society can make true statements about foreign human experiences. Relativism tends to be asserted in societies undergoing rapid political, social, or technological change, but this alone does not prove it is false (or true).

Human thought need not be entirely contingent: the circumstances in which knowledge is discovered are not the same thing as its content. That $2 + 2 = 4$ is not dependent on how you learned to add, or even whether the sum was first computed with sticks, fingers, or a hand calculator. An objectification of nature derived from capitalist industrial society may have been necessary for a scientific concept like the periodic table; does this mean that the resulting principles are false?

At this level of generality, parallels between scientific knowledge and social structures or cultural practices are not very helpful. As has been suggested above, Leonardo and Machiavelli seem to have outlined the approach of modern science before the social practices associated with it were fully in evidence. On the other hand, both Neoplatonist and Aristotelian modes of thought continued to exercise enormous power throughout the Christian Middle Ages despite the disappearance of the polytheistic religious matrix within which ancient science emerged.

A good example of the difference between the motives or causes of a scientific theory and its truth is provided by the principles of statistics used to verify whether the evidence supporting a scientific hypothesis could have occurred by chance. Statistical tests of significance were first worked out in detail by R. A. Fisher, a geneticist interested in extending the principles of plant domestication to human populations. Fisher developed these statistical methods in the service of positive eugenics, with the goal of halting what he perceived to be the degeneration of the human species' vigor under the conditions of industrial society.[5]

Fisher was among those for whom science proves that human intelligence is largely inherited. Many critics of this position (including Kamin, Chomsky, Gould, Bowles, and Gintis) view the eugenics movement as a factor supporting the Nazi policy of genocide, yet they use Fisher's statistical methods to show that IQ is not transmitted genetically. No one has ever claimed that these statistical methods are biased merely because they were first developed in the service of a highly controversial extension of agronomy to human populations. Scientific principles may be incomplete, but their truth can hardly be denied merely because of the type of society in which they emerged or the personal motives of the theorist.

The claim that science is colored by intrinsic bias might, however, have a different meaning. Perhaps relativism infects scientific theories by introducing a consistent value commitment to specific concepts and findings. To see whether the content of a theory is biased, it is necessary to change the method of analysis by considering the way similar ideas have been used in different times and places. To this end, I will consider two theoretical issues that lie at the heart of any theory of human nature: the origin of species (or the nature of evolutionary change) and the foundation of society.

In both areas, contemporary scientific theories have been vigorously attacked as ideological. It has been charged that some recent biological theories (evolutionary gradualism, sociobiology) are inherently conservative if not reactionary, while others (cultural determinism, punctuated equilibrium modes of evolution) have been dismissed as reflections of left-wing bias. Historical evidence shows, however, that similar theoretical concepts have been associated with the most diverse political, ideological, and religious perspectives.

This procedure may seem unusual, since contemporary scientific theories are often said to have few antecedents in antiquity worth considering.[6] More often than not, as discussions of modern physics suggest, the opposite is the case.[7] While Darwinian biology was contrary to the religious doctrines that had prevailed in the seventeenth and eighteenth centuries, evolutionary theories clearly existed in other times and places. We know, for example, that some ancient Greek thinkers explored the origin, persistence, and modification of species in ways that are directly relevant to the issues of modern science.

To be sure, these ancients should not be treated as if they were linear precursors in the development of modernity. Stephen Jay Gould was quite correct, on methodological grounds, to "reject an approach to the history of science that rapes the past for seeds and harbingers of later views; such a perspective only makes sense within the abandoned faith that science progresses by accumulation towards absolute truth."[8] But one cannot go to the other extreme and claim that all prior thought is irrelevant without implying that each epoch is unique. Such extreme historical relativism would assume precisely the point at issue.

The Origin of Species and Human History: The Political Implications of Punctuated Equilibria Theory

Contemporary biology provides several good examples of the claim that scientific theories are vitiated by political bias. According to the conventional interpretation of Darwin's theory of natural selection, species originate and evolve by slow, incremental changes. A number of scholars have attacked this evolutionary theory, asserting that it is biased in favor of conservatism or reformism. Stephen Jay Gould and Niles Eldredge (among others) have proposed an alternative theory, known as punctuated equilibria, according to which evolution is marked by brief periods of enormous change followed by long periods of stability.

In presenting the model of punctuated equilibria, Gould and Eldredge refer to the fact that one of them learned Marxism "at his father's knee."[9] Such an explanation of the origin of animals seems biased in favor of fundamental political change if not violent revolution. Not surprisingly, therefore, the concept of punctuated or discontinuous evolution has been associated with Marx's theory of history and accused of left-wing bias.

It seems intuitively obvious that theories of biological evolution should be used as metaphors for human history and political change. The evidence from the history of ideas shows, however, that this is not the case. Theories of either gradual or discontinuous natural change have been associated with the most diverse theories of human history and with varied political preferences (Table 6.1).[10]

Descriptions of biological or natural change as a series of catastrophes are quite old: the image of a primeval flood (if not of Noah's Ark) is found in many cultural traditions outside the Hebrew Bible, associated with varied religious beliefs that could hardly be called revolutionary. Even if a survey is limited to scientific and philosophic accounts of biology and human history, theories similar to the Gould–Eldredge model seem to be associated with the diverse political preferences.

TABLE 6.1
Concepts of Change in Nature and Human History

	Biological Evolution (Origin of Species)	
Human History (Origin of Civilization)	Gradual change	Discontinuous change
Gradual change	Aristotle Sumner	Lucretius
Discontinuous change	Rousseau	Hobbes Marx

Empedocles and the Epicureans. In Greek antiquity, Empedocles developed a theory of evolution that, not unlike Darwinian biology, postulates a process of random variation and a kind of natural selection.[11] His understanding of evolution resembled the Gould–Eldredge model of punctuated equilibria: periods of rapid change, in which random events produce a wide variety of forms, lead to the selection of types that tend to be conserved over long periods.[12]

Although Empedocles himself did not leave behind a comprehensive treatment of politics, similar biological and political ideas were expressed by the Epicureans, most notably in Lucretius's *De Rerum Natura* (On the Nature of Things).[13] After presenting an account of organic and human evolution that bears many structural similarities to that of Empedocles (5.772–1457),[14] Lucretius describes human history as slow and continual change (5.925–1457).[15] Substantively, Lucretius and other Epicureans use the gradual emergence of human political institutions to demonstrate that justice is not a thing in itself but rather is an agreement based on the mutual interests of individuals. Since standards of justice vary from one time and place to another, the natural end or goal of human life is not to be found in politics. Lucretius thus follows Epicurus in insisting that the prudent man will avoid political commitment as well as all other activities that risk his calm and contemplation. In this example, a biological theory like that of punctuated equilibria was combined with a gradualist view of human history and hostility to all forms of political commitment—that is, to human values that are exactly the opposite of those found in Marxism.

Hobbes. Although Hobbes was a secular thinker who does not seem to accept the biblical account of creation as literally true, he does not focus on evolutionary change in humans or other animals. Hobbes combines a view of the natural fixity or stability of animal species with a highly discontinuous interpretation of human history. Perhaps even more important, Hobbes sees his own principles as having potentially enormous or "revolutionary" effects, since their general acceptance will make possible an "eternal" political community that channels violent conflict into peaceful social cooperation.[16]

Like Lucretius, Hobbes holds that there is no natural fairness or justice without the enforcement of man-made laws; like the Aristotelian tradition challenged by Lucretius, Hobbes praises political life and presents a science of politics designed to provide concrete guidance for political leaders. But Hobbes differs from these predecessors in his optimism for the future, based on the assumption that natural science and technology could conquer natural necessity, material scarcity, and political conflict. Even more important than the rate of change for Hobbes is its direction: he foresees historical progress based on scientific knowledge.

Marx. Although it has frequently been said that Marx offered to dedicate *Das Kapital* to Darwin,[17] Marx's understanding of evolution follows Hegel in claiming that our species is unique (in contrast to Darwin's denial of a sharp gulf between humans and other animals).[18] Humans have essentially created themselves through the process of free, conscious labor.

Human history—"nature developing into man"[19]—is punctuated by a sequence of revolutionary upheavals followed by more gradual development within each established form of social relations.[20] Marx's view of human origins thus reflects the same kind of discontinuity as his theory of history. On this basis, he develops the substantive political principles now associated with communism: progress through a revolutionary political movement toward the definitive liberation of all mankind in abundance and justice.[21]

If the theories of Lucretius, Hobbes, and Marx are contrasted, we find that the concept of discontinuous, or punctuated, biological change has been associated with diverse views of human history and politics. For the ancient Epicureans, change in human history was gradual, and the goal was a retreat from political life; for both Hobbes and Marx, the same evolutionary concept was associated with discontinuous historical change but combined with opposed views of politics. The concept of punctuational change is, in itself, not biased toward specific values or social goals.

Natural Selection and Human History: The Political Implications of Evolutionary Gradualism

A similar diversity of political implications has been associated with the concept of evolution as a process of gradual change and adaptation. For many thinkers, natural events have changed slowly if at all: either animals have always been as they are today or their evolution has been the slow and progressive modification described by Darwin himself in *Origin of Species*. A comparison of three thinkers—one ancient (Aristotle) and two moderns, the former anticipating Darwin (Rousseau) and the latter influenced by him (Sumner)—illustrate this variety of implications.

Aristotle. Aristotle wrote extensively and descriptively about animals, but his primary topic was the analysis of form and function—what today is called adaptation, ontogeny, and systematics—rather than evolution.[22] Aristotle's main conclusions can be stated in modern terms: first, some but not all resemblances between parents and their offspring can be inherited; second, some of these similarities are due to developmental patterns rather than to the physical state of parents at procreation; and third, the transmission of recessive traits would be impossible without a material basis of inheritance.[23]

The functional integrity of organisms and the regularity of their development led Aristotle to reject both direct somatic inheritance and Empedocles' account of the origin of species in a chaotic epoch in which "isolated parts" were "combined into animals."[24] Aristotle's tendency to treat existing animals as given therefore does not presume species have been fixed for all time. Rather, his view of change can be compared to gradualist models of evolution, particularly since Aristotle flatly denies the eternity of living beings and stresses the mechanisms of small but continual changes both within and between species (*Physics*, 1.6–9, esp. 8.191b; *Generation and Corruption*, 1.3–5; *History of Animals*, 8.1.588b; *Parts of Animals*, 1.1.639b—and, for humans, *Politics*, 2.8.1269a3–6).

The link between Aristotle's biology and his political theory is symbolized by the term *zoon politikon* ("political animal"), which he uses to describe the human species.[25] Because his concept of change in human history and biology is similar, Aristotle is perhaps the most consistent gradualist in the Western intellectual tradition. Substantively, Aristotle's political principles include the acceptance of moderate socioeconomic inequality and (at least for some individuals) even slavery, a preference for the mixed regime in an ancient city-state, the rule of law, the rejection of militarism (including tacit criticism of expansionist empires like that of Alexander the Great), and insistence that political activism is lower in status than scientific inquiry.[26] The virtues are "by nature" and can generally be described as means between extremes; the virtuous human being has a character exemplified by rationality, moderation, and concern for the social consequences of individual behavior.

Whatever the differences between one society and another, natural justice exists and forms the basis of political judgment (*Nicomachean Ethics*,
5.7). While Aristotle was clearly not in favor of tyranny, he could hardly
be called a democrat in the modern sense; he stressed the importance of
moderation and the idea of a mixed, limited regime dominated by a class
of leisured, well-educated, and virtuous gentlemen. Political participation,
particularly in a community that permits "ruling and being ruled in turn," is
a proper and excellent exercise of the natural faculties of a human being. As
we will see, gradualist principles in biological evolution and human history
have sometimes been associated with quite different political principles.

Rousseau. Writing a century before Darwin's *Origin of Species*, Rousseau took a gradualist view of evolutionary change.[27] Despite his evolutionary gradualism, Rousseau's view of human history resembles punctuated
equilibria rather than continuous infinitesimal changes.[28] Arguing that environmental factors like climate influence politics as well as biology, Rousseau therefore combined a gradualist theory of biological change with a
punctuational view of human history in formulating one of the most radical
political teachings in the Western tradition.[29]

Rousseau's substantive conclusions were anathema to the ancien régime
(which banned his works and condemned his teaching): the only political system worthy of obedience is a community of relatively equal citizens
who are free to enact the laws and elect the governing authorities annually.
For Rousseau, therefore, gradual processes in biology did not entail either
historical gradualism or deference to constituted political authority. Few
political theorists have articulated a clearer account of human origins in
a natural process of gradual change, yet few have been more profoundly
radical in challenging existing political institutions.[30]

William Graham Sumner. After Darwin, the gradualist theory of evolution was given a quite different political application by the movement
known as social Darwinism. Broadly speaking, these thinkers emphasized
the notion of natural selection as a struggle culminating in the "survival of
the fittest" and translated it directly into the sphere of human economics
and politics. In so doing, they developed a powerful argument in favor of
laissez-faire economics and the unhindered development of wealth.

This view is well illustrated by William Graham Sumner's *What the
Social Classes Owe Each Other*,[31] which explicitly adopts a gradualist
evolutionary approach and uses it to explain human history.[32] Social and
economic life is inevitably competitive, leading inevitably to the difference
between rich and poor; even cooperation is based on the self-interest that
comes from a "greater and greater control over Nature."[33] It is a mistake

to try to manipulate this outcome because redistribution of wealth from rich to poor is always harmful.[34]

Many contemporaries have cited Sumner's argument as proof that evolutionary gradualism inevitably has reactionary political implications. But at about the same time, Kropotkin was using Darwin's principles to conclude that cooperation, not competition, is a law of nature.[35] And, of course, not a few contemporary biologists have adopted gradualist theories of natural selection without coming to Sumner's conclusions.

The Origins of Society: Methodological Individualism versus the Social Construction of Reality

The foregoing survey of evolutionary theories, while highly selective, shows that thinkers with similar views of evolutionary or historical change can espouse the most diverse political conclusions. It does not seem that scientific concepts of the origin of species are narrowly bound to the social conditions in which they were formulated. This characteristic is not limited to issue of evolutionary discontinuity or gradualism, for it can also be found in theories explaining the origins of society.

In recent years, the version of evolutionary theory known as sociobiology has been attacked as a mask for capitalist or bourgeois economic interests. For sociobiologists, because natural selection operates primarily on individuals, social cooperation can be understood as a means of pursuing the self-interest of the individual organism—or, at least, of the organism's genes. In the eyes of its critics, this imposes on nature the cost–benefit analysis of entepreneurs in a market economy.[36]

Those hostile to sociobiology have argued that society has properties that cannot be derived from the interests of individuals. As it relates to humans, this view explains culture as the product of conventions or norms that cannot be reduced to the cost–benefit calculus of individual selfishness. Some social scientists have spoken of the "autonomy" of culture and society, while others described this view as the "social construction of reality."

Both of these explanations of the origin of human society have been attacked as ideologically biased. For the left, sociobiology is a form of "methodological individualism" justifying capitalism. For the right, critics of sociobiology are merely extending Marxist commitments into science.

Such charges imply that each theory of the origin of society should be consistently associated with the corresponding political principles. Once again, the evidence shows this is not the case (Table 6.2).

TABLE 6.2
The Origins of Society and Varieties of Political or Social Theory

Type of Political or Social Theory	Origins of Society	
	Methodological Individualism	Social Construction of Reality
Ancient thinkers	Antiphon the Sophist Thrasymachus Gorgias	Socrates Plato Aristotle
Modern political theories	Hobbes Locke Rousseau	Machiavelli Montesquieu Marx
Nineteenth-century Darwinists	Sumner	Kropotkin
Contemporary social science	Economists Game theorists	Sociologists Historians
Contemporary biology	Sociobiologists	Ecologists

Individualism and Cost–Benefit Theories of Cooperation. Sociobiologists present an evolutionary explanation of social behavior among animals, based on the principle of natural selection. Organisms are more likely to reproduce if the benefits of their behaviors are more than their costs. Indiscriminate altruism (e.g., dying to assist those who are unrelated and therefore genetic competitors) is not generally observed among animals. According to rigorous supporters of the theory, the measure of natural selection can concern only the survival and reproduction of individual descendents, not the benefit of the group.

In dealing with humans, the claim that a complete scientific theory can derive all complex social patterns from the costs and benefits of individuals has been called methodological individualism. In the social sciences, this approach is used by many economists, game theorists, and rational-choice theorists. Similar theories have, however, been developed by theorists in different cultures and with widely varied preferences and goals.[37]

The premises of methodological individualism had already been developed by the Sophists in ancient Greece, as is evident in Plato's *Republic*, Antiphon's *On Truth*, Aristotle's reference to Lycophron, and other textual sources.[38] Among the ancients, these individualistic premises were associated with quite diverse political preferences and values. For Thrasymachus or Gorgias, the best human life is that of the successful tyrant, who uses knowledge of the conventional origins of law in his private self-interest. For Antiphon as well as for Epicureans like Lucretius, the private or philosophic life is the best. And for the author of a text known as the Anonymous Iamblichi, the conclusion is that education is essential to produce civic virtue.

In modern times, the theory that society emerges as the most efficient way for individuals to balance the costs and benefits of achieving private goals was developed by Hobbes, Locke, and Rousseau, among others. Yet these modern philosophers derive a variety of political prescriptions from similar individualistic premises: absolute authority and strong monarchy (Hobbes); a division between the legislative, executive, and "federative" powers (Locke); the assembled people, as in an ancient city-state, as the only legitimate sovereign (Rousseau).

Methodological individualism has survived in a variety of disciplines within the social sciences. In psychology, for example, a tradition has been traced from Hobbes and Locke to Watson, Hull, and the behaviorism of B. F. Skinner.[39] The intellectual history of modern economics can also be traced to similar roots through the influence of Sir William Petty (a seventeenth-century follower of Hobbes), Townsend (whose critique of the poor laws was based on an argument akin to the survival of the fittest), and Malthus. When Darwin developed the principle of natural selection to explain evolution, he was explicitly building on a tradition of individualism that had been applied to human behavior for centuries.

The historical evidence shows the variety of positions derived from individualistic premises. It may seem intuitively obvious that the use of Darwinian biology to explain human behavior reflects capitalist ideology because both share the individualist premises of Locke's political theory. In fact, the assumption that social communities are the products of individual selfishness has been associated with quite diverse political and social values, ranging from Thrasymachus's praise of tyranny and the Epicurean condemnation of all political life to Rousseau's communitarian hostility to the large-scale commercial societies of modern times. It is hard to see how a theoretical concept that has been combined with such varied political and social conclusions is inherently biased.

Natural Sociability and the Autonomy of "Social Facts" Confronting methodological individualism is the view, variously described as "the social construction of reality" or the autonomy of culture, that the very notion of the individual is a product of historical conditions. Although also criticized as ideologically biased, theories based on the premise of natural sociability are as varied in origins and political implications as are those of methodological individualists. Plato, Aristotle, and the entire Socratic tradition denied that societies are entirely conventional or artificial; as Aristotle put it, the human being is a "political animal."[40] In the Middle Ages, similar views were associated not only with Thomist thought but with other theological and philosophic positions in the Islamic and Jewish as well as Christian traditions. Yet these theories of human sociability have been associated with diverse religious teachings (paganism and three distinct forms of monotheism) and directed to diverse aims (from the primacy of the philosophic life to the support of political action, the pope, or the Holy Roman Emperor).

In modern times as well, a variety of positions have been associated with hostility to individualist premises. For sociologists like Durkheim and Weber, "social facts" are in their nature irreducible to individual choices; in this view, social systems have properties that are sui generis. Marx and the Marxist tradition share this premise and use it to show that the existence of the individual as a social phenomenon is the result of capitalism and bourgeois ideology. Although conventional wisdom contrasts the revolutionary left of Marxism with the supposedly conservative sociological theories of Weber and Parsons, both share similar models of the origin of society.

Even in applications of Darwinian biology, one can find theorists who deny the individualist premise. Kropotkin long ago challenged Darwin's interpretation of nature as basically competitive; more recently, Adolph Portmann emphasized that many animals are "social beings" by their very nature. In contemporary debates on natural selection, some scholars have taken the side of so-called group selection, emphasizing phenomena that they claim cannot be understood on individualist premises.[41]

Premises concerning the primacy of the individual or the society have thus been in dialectic relationship to each other throughout the history of Western thought. The diversity of contexts in which the same basic theoretical model has been developed should give us pause about the relativists' claim that ideas have "inherent" political bias due either to their social origin or their content. Since concepts of history (discontinuous vs. gradual change) and the origins of society (the primacy of society vs. individualism) can be combined with other theoretical principles in almost every possible way, the appropriate question is one of scientific evidence.

Structuralism and Scientific Theory

The biases attributed to scientific theories do not explain either the diversity of thinkers who adopt similar concepts nor the variety of uses to which these ideas are put. The relativist or intuitionist position has been

TABLE 6.3
Some Structural Oppositions in Science and Philosophy

Question	Positive	Negative
Do universals exist in nature?	Nominalist	Realist
Is "free will" a cause?	Voluntarist	Determinist
Is knowledge true in all times and places?	Absolutist	Relativist
Are visible things the primary reality?	Materialist	Idealist
Is there a single principle of all things?	Monist	Pluralist
Are theories approximations needing modification on the basis of practice?	Pragmatist	Doctrinaire
Do species originate in discontinuous process?	Punctuational equilibria	Evolutionary gradualism
Are societies explained by individual choices?	Methodological individualism	Social construction of reality

described as holding that "there will always in principle be indefinitely many, genuinely different ways of translating any body of discourse from one language to another."[42] On the contrary, it would seem that in science, as in myth, there is a finite set of structural or conceptual components, which are combined and recombined in many ways. Claude Lévi-Strauss has shown that such a structuralist approach helps us to understand many myths and social practices that have long puzzled anthropologists.[43] Perhaps the same process of combining and recombining distinct elements lies at the root of diverse scientific theories.

In the history of science and philosophy, a number of fundamental issues can be identified. Each of these basic questions tends to be answered in two or three ways, so theorists can typically be classified in one or another opposed camp (much as we have seen with regard to the evolution of species and origin of human societies). At the risk of oversimplification, some of these basic oppositions are summarized in Table 6.3.

From a structuralist perspective, each distinct theory—whether in the natural or the social sciences—can be understood as a combination and permutation of the varied concepts listed in Table 6.3. Any one of these concepts may be developed for the first time under specific social and political circumstances; at first, the proponents of the concept may well present it for narrowly self-interested reasons. But if the overall domain of science concerns these underlying structures, the truth of concepts would have little to do with the historical conditions of their discovery or use.

Instead, science could be seen as the process of discovering or defining more precisely the domain to which a given concept appropriately belongs.

Quite typically, scientific debates reveal that both sides have an element of truth. Is light a wave or a particle? Clearly both. Is human behavior due to nature or to nurture? Clearly both. Once we move beyond the proclaimed impossibility of human knowledge,[44] the charges and countercharges of political or cultural bias can be replaced by instructive dialogue.

PART THREE

FROM FACT TO VALUE

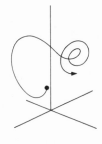

7. Scientific Value Relativism: Locke, Hume, and Value-Free Science

The problems confronting our civilization seem related to the triumph of modern science, yet we are told scientific knowledge cannot be the foundation of our goals and values. The evidence that knowledge of reality is possible, presented in part 2, does not in itself provide standards of moral and political judgment. Why do so many contemporaries believe that facts are so sharply divorced from principles of right and wrong?

This value-free character of contemporary natural science has not usually been problematic: questions about the fruits of science are typically directed to the way society *uses* science rather than to the scientific project itself.[1] Today, however, the awesome policy implications of scientific and technological discoveries make it difficult if not impossible to ignore the problem of *scientific value relativism*.[2]

Three questions are the focus of the following chapters. First, how did the divorce between scientific knowledge and value judgment originate in the thought of Locke, Hume, and the positivist tradition (chap. 7)? Second, why does recent scientific evidence contradict the premises of scientific value relativism (chap. 8)? And finally, what are the outlines of a scientific understanding that integrates nature and nurture, making it possible to bridge the gap between fact and value in a reasonable manner akin to that of ancient science (chap. 9)?

The Problem of *Is* and *Ought* in Modern Science

Could a deeper understanding of science itself help lead us out of the confusions of modern and postmodern thought? The reasons that supposedly divorce science from human values need to be reconsidered in the light of modern science itself. When we do so, it will become evident that the so-called naturalistic fallacy is itself a fallacy.

None of the defining characteristics of modern science or technology distinguishes fundamentally between animate and inanimate objects of knowledge—or, within the animate world, between humans and other living things. The modern approach implies the existence of *social* or *human* sciences as well as *natural*, *physical*, and *biological* ones.[3] But since humans are themselves the creators or researchers in the study of humans, how can the same beings be both the subjects and the objects of a true science? Are there scientific reasons why humans do (or do not) study science?

These questions amount to asking what ends or purposes are served by studying humans from a scientific perspective. As a noted psychologist put it recently: "Many scientists shy away from exploring values, because they believe there is no way of getting from *is* to *ought*; from knowing how the body and the mind grow to concluding how they should grow. . . . We are all still rather inept at justifying the leap from empirical facts to ethical convictions."[4] More simply, can science tell us what humans should do?

For many contemporaries (among nonscientists as well as scientists), the question was settled long ago: modern scientific knowledge cannot establish the "ends" or purposes for judging human life. This self-imposed limitation is reflected in the assumption that there is a fundamental difference or gulf between facts and values, with science focused on the knowledge of the former. In principle, if not in practice, scientific methods and theories are thought to be value-free. To use a technical term of some philosophers, it is appropriate to adopt the position of scientific value relativism.

This view is not silly. How could things be good merely because they happen as a result of natural necessity? A stone falls: is it desirable? An infant is born with a genetic abnormality, suffers, and dies: who would claim that the natural causes make the infant's suffering and death preferable to a healthy and happy life? Little wonder that any attempt to deduce values from facts is said to commit the naturalistic fallacy. But could that be the whole story?

Before we can answer this question, it is necessary to understand the radical distinction between *Is* and *Ought* so widely accepted today. There seem to be three main arguments against the ancient view that reasoning can discover, in human nature, the foundations of ethics and morality.

First, cultural variability in standards of right and wrong makes it hard to see why any one set of beliefs or practices is more natural than any other; throughout history, virtually every imaginable action, from cannibalism to murder, has been considered legitimate or praiseworthy by some human society. Second, the theory of knowledge seems to show that the human mind cannot discover universal truths; even if there were a natural foundation to justice and virtue, we could not know it with the degree of certainty required to make it an obligation or duty. And finally, logic teaches that it is irrational or self-contradictory even to seek a natural foundation for morality and values; only logical or mathematical analysis provides a way of demonstrating truths—and such an analysis shows that values are little more than individual and noncommunicable preferences.

To explore how the modern approach to these issues originated, the thought of Locke, Hume, and Kant will be examined and related to a statement of the twentieth-century positivist view by Arnold Brecht. When carefully reconsidered, these arguments do not establish the fact–value dichotomy in a convincing way. Indeed, a close reading of Hume reveals that he did not adopt the sharp distinction between *Is* and *Ought* widely attributed to him today. As a result, the premises of the value-free approach to science need to be reassessed in the light of recent scientific discoveries.

Locke and the Origins of Value-Free Science

Although John Locke greatly influenced the modern tradition through his political ideas, which shaped the American constitutional tradition,[5] his theory of knowledge had at least as much of an effect on our thought. Locke's *Essay Concerning Human Understanding* begins from the question at issue: how do humans know the grounds of moral obligation? He claims there are three fundamentally different ways to answer questions of morality, duty, and right:

That men should keep their compacts is certainly a great and undeniable rule in morality. But yet, if a Christian, who has the view of happiness and misery in another life, be asked why a man must keep his word, he will give this as a reason:—Because God, who has the power of eternal life and death, requires it of us. But if a Hobbist be asked why? he will answer:—Because the public requires it, and the Leviathan will punish you if you do not. And if one of the old philosophers had been asked, he would have answered:—Because it was dishonest, below the dignity of a man, and opposite to virtue, the highest perfection of human nature, to do otherwise.[6]

For Locke, the "fundamental rules of morality" could come from one of three sources: Christianity (religious belief justifies moral obligation), a "Hobbist" or modern science (which explains morality in terms of the conventions and laws of each society), or the "old philosophers" (a rational assessment, characteristic of such ancient thinkers and scientists as Plato and Aristotle, that derives morals from "the highest perfection of human nature"). In short, Locke begins from the three basic ways of knowing that I discussed in chapter 1.

Of these three explanations for morality, Locke's own position seems akin to that of Hobbes.[7] For Locke, as for Hobbes, moral rules or duties are the result of *human* convention or law, to be explained as the result of either social custom and habit ("the public requires it") or punishment by the state ("the Leviathan will punish you"). Like Hobbes, Locke's understanding of science focuses on the means individuals choose to seek private ends; he does not accept the ancient view that the end or virtue of a human life can be understood as given by nature, or the Christian view that human knowledge and morality can be derived from divine inspiration or faith: "practical principles" are not "innate and imprinted in our minds immediately by the hand of God" (*Essay*, 1.2.5; vol. 1, p. 69).

Locke's attack on innate ideas has two implications: first, God did not communicate universal standards of behavior to all humans; and second, such rules or norms cannot be derived from nature by the philosopher or scientist.[8] Every known norm or rule of justice—most specifically, those rules of conduct considered "natural" by the philosophic tradition—is violated by entire societies. Locke's principal example (which will be examined in detail in the next chapter) is parental care of children, a seemingly innate obligation that is violated not only by individuals engaging in child abuse but also by the cultural practice of infanticide found in many societies. Moral rules cannot be innate or natural.[9]

In sum, for Locke, the premises of both religion and ancient philosophy are inconsistent with the observed behavior of humans. The "old philosophers" had taught that justice and the other virtues were "according to nature" and could be discovered by reasoning about the ends or essence of human life.[10] Such a view presupposes not only some innate knowledge of right and wrong, but punishment by nature for a failure to obey the rule— and no such punishment exists. Christianity teaches that there are divine commands and punishments, but revelation is obviously not the same thing as a natural instinct. The natives discovered in Africa or America by Western explorers had no way of knowing of Christ, and their religious beliefs justified an immense variety of moral standards and practices.

Locke does not deny that scientific knowledge of human behavior is possible. When studied objectively, however, humans are found to have "natural inclinations" that are essentially *contrary* to morality.[11] According to Locke's epistemology, which came to dominate Anglo-Saxon science, if not modern science more generally, rules of conduct—and therewith ethical values—come from education, culture, or tradition, and ultimately from positive law (*Essay*, 1.2.23; vol. 1, p. 87–88).

The desires of humans are given by nature, whereas the objects of the desires, and hence both the goals of human endeavor and the duties or laws on which morality rests, depend on convention and learning. For Locke, as for other modern thinkers, the science of human nature cannot be reasoned dialogue concerning the best human life or the best political regime, as it had been for the "old philosophers" in the tradition going back to Socrates.[12]

Humans have, in this view, "natural rights" such as the right to "life, liberty, and estate." By nature, each individual is free and equal.[13] But this natural right does not indicate the proper ends or values for human life; rather, it suggests that science serves to discover the most efficient *means* that individuals might use to achieve their ends. At most, the Lockean science—like the principles of Hobbes before him—points to behaviors that are inconsistent with probable survival and therefore highly costly.

Locke therefore establishes the typically modern tradition that science must be value-free with regard to the highest questions of right and wrong. As Beckstrom put it, modern science is like the airline ticket agency in suggesting efficient or economical routes to achieve a destination but not suggesting where one should go. Or to use the ironic phrase of Leo Strauss, modern science engages in "retail sanity and wholesale madness."[14] One of the most obvious limitations of modern science is its unwillingness or inability to treat the goals of human life (values) as subjects for scientific inquiry.

Hume and the Naturalistic Fallacy

For many contemporary students of philosophy, Hume rather than Locke best formulated the scientific understanding of what humans can and cannot know.[15] Hume's skepticism, notably with regard to the limited knowledge of causal relationships, is widely viewed as unassailable. Since the concept of the naturalistic fallacy is often traced to Hume, careful consideration of his epistemology and its relationship to the questions of morality, justice, and obligation is essential.

For Hume, the study of human understanding is founded on a basic principle (derived from Locke but interpreted in the light of subsequent thought) summarized as follows: "*all our ideas are copied from our impressions.*"[16] This principle confirms Locke's insistence that there are no innate ideas, but it also provides the foundation of Hume's more fundamental challenge to demonstrable or absolute knowledge of causal relationships. What humans call a cause is merely a conventional or habitual relationship between ideas.[17]

In principle, therefore, one can never know whether future events will correspond to those of past experience.[18] Because essences are unknowable, all assertions of causality are provisional; in observing and judging the world, humans can never go beyond relationships that are probabilities or possibilities. Even "the supposition, *that the future resembles the past, is not founded on arguments of any kind, but is derived entirely from habit*" (*Treatise*, 1.3.12; p. 123). Absolute certainty is inaccessible to human understanding. All knowledge of nature is hypothetical.

Hume's theory of knowledge thus undermines all claims to have discovered any absolute definition of right and wrong. His reasoning not only proves "that morality consists not in any relations that are the objects of science" but also that "it consists not in any *matter of fact*, which can be discovered by the understanding" (*Treatise*, 3.1.1; p. 422).

Hume's distinction between morality and science or fact is explicitly associated with the danger of moving from "the usual copulations of propositions, *is*, and *is not*," to propositions "connected with an *ought*, or an *ought not*." Hume explains this danger: "For as this *ought*, or *ought not*, expresses some new relation or affirmation, it is necessary that it should be observed and explained; and at the same time that a reason should be given, for what seems altogether inconceivable, how this new relation can be a deduction from others, which are entirely different from it" (ibid., p. 423).[19]

This distinction between Is and Ought has become one of the classic arguments for a value-free science. But although Hume's probabilistic approach to the sciences has much to recommend it,[20] it differs from the understanding of the fact–value dichotomy now prevalent. For many social scientists today, Hume's distinction between Is and Ought is taken out of context and treated as the logical foundation of the scientific method.

Hume himself rejects the attempt to derive scientific method from logic *as distinct from* the study of human cognition because the very idea of "necessity" underlying analytical logic or mathematics is itself derived from experience.[21] Whereas many contemporaries derive the methods of scientific knowledge from logic, Hume derived the possibility of logic from the scientific understanding of human knowledge.

As we will see, Hume was actually far from endorsing the view of science and morality often attributed to him. Indeed, as several scholars have recently suggested, Humean principles may be closer to Aristotelian naturalism than to contemporary relativism.[22] Why, then, has the fact–value dichotomy been so widely adopted by our contemporaries, particularly in the social sciences? To answer this question, it is necessary to consider the argument for scientific value relativism set forth by those called logical positivists.

Kant, Logical Positivism, and the
Fact–Value Dichotomy

Today, the presumed gulf between facts and values dominates the application of science to human affairs.[23] The roots of this view can be traced to Locke, who classified the sciences into three branches called "physics" or "natural philosophy," "practics" or "the science of ethics," and "semiotics" or "logic."[24] Although this threefold classification has become the implicit framework for understanding modern science and its relation to human society,[25] neither Locke nor Hume went as far as our contemporaries in asserting that the logical distinction between fact and value must govern both social and natural science.

The current view that logic is the principal foundation of both the scientific method and moral reasoning was greatly shaped by Immanuel Kant's response to the skeptical tradition of Locke and Hume. To save reason and moral judgment from the challenge that all thought is merely convention or habit, Kant accepted the impossibility of knowledge about the essence or inner being of things—and tried to turn this limitation into the basis for moral certainty.

By means of the *Critique of Pure Reason*, as his famous work is entitled, Kant concludes that all experience is known to us only through sense experience of "phenomena," whereas the "noumenon"—the thing in itself— remains utterly inaccessible. There is one critical exception, however: the human will, the noumenal realm of human reason, judgment, and moral choice. While Kantian metaphysics influenced methods in the natural sciences,[26] its sharp distinction between the study of human morality and the scientific analysis of other things was probably even more important.

Kant's analytical search for the fundamental "antinomies" on which thought must rest seems directly opposite to the empiricism of Locke and Hume, yet both led to a divorce between fact and value. Where Locke derived knowledge and obligation from education or custom, and Hume taught that duty and virtue is the result of a "moral sense" based on *feelings* or *emotions*, Kant dismisses such materially determined factors as phenomena irrelevant to the determination of the "pure" will.[27] Any admixture of interest or feeling derived from our animal nature is, for Kant, inconsistent with the purely noumenal definition of duty.

Quite obviously, the argument is at bottom logical or metaphysical rather than empirical or descriptive. Kant claims that, as a matter of principle, morality cannot be deduced from "particular attributes of human nature": "every empirical element is not only quite incapable of aiding the principle of morality, but is even highly prejudicial to the purity of morals."[28] In the Kantian view, it is impossible to have clarity about either empirical science or ethical norms without first accepting the antinomy of noumenon and phenomenon, placing morality and duty on the side of the noumenon. The facts have nothing to do with values.

This aspect of neo-Kantian reasoning is so widely shared in contemporary social science that it is almost invisible.[29] Its status as one of the principal foundations of the fact–value dichotomy, and therewith the project of a value-free science of nature and human nature, is evident from the resistance to studies of the material "causes" of human behavior. When Marxists attack the fact–value dichotomy as bourgeois ideology, or a psychologist like Skinner rejects it because "values" are merely positively reinforcing stimuli or social actions, their critics echo Kant's complaint that such scientific explanations would rob humans of their "freedom."[30]

Kant's solution to the problem of morality, the Categorical Imperative, takes the form of a purely logical or rational procedure. The only ground for duty is to act so that the maxim or rule governing one's will could be a universal moral rule. I may wish to rob a weak old man, but I cannot rationally adopt the maxim that "robbery is legitimate" lest my gains be stolen in turn; the act is wrong. In contrast, if I wish to help the old man cross the street, the act can be deduced from a general rule that "the healthy should help the aged and infirm"; this act can be done with a good or moral will out of pure duty.

By deriving absolute ethical principles from the "synthetic a priori" (a logical category derived from the study of pure reason), Kant could claim to have established a moral rule that was insulated from the skeptical empiricism of Locke and Hume. Ironically, however, Kant's emphasis on the priority of logic ultimately had the opposite effect, providing the foundation for scientific value relativism. To see how the Kantian focus on logic and method came to justify the relativism so pervasive today, it is useful to consider a statement of the fact–value dichotomy by a leading twentieth-century logical positivist.

Arnold Brecht's *Political Theory* represents the so-called Vienna Circle that dominated much of the discussion of philosophy and method in American social science during the middle of the twentieth century.[31] For Brecht, the problem in contemporary thought resulted from "the rise of the theoretical opinion that no scientific choice between ultimate values can be made" (p. 9). A thorough understanding of the reason for this depends on logic, since "logical reasoning is accepted as *full* proof by Scientific Method when, and only when, it is strictly analytic" (p. 55).

This conception of logic buids on the Kantian distinction between analytic and synthetic deductions: the latter "adds something to the meaning of a given term or proposition," whereas the former "merely makes explicit what is implied in that meaning" (ibid.) The premises or postulates of logic or mathematics "may be grounded on the belief that they reflect existential reality or self-evident truth" or "based on sheer expediency and even on arbitrary choice" (p. 56). Whatever these premises, as long as an argument is "strictly analytic in drawing implications," its results can be "absolutely and demonstrably true" (p. 57).

For Brecht, it necessarily follows that there is a logical gulf between statements of what *Is* (fact) and of what *Ought* to be (value).[32] In human affairs, the questions of ends or goals need to be viewed as moral, religious, or personal values that are distinct from knowledge of fact. Methodological analysis establishes scientific value relativism, a doctrine that science can explore the implications of value choices but never provide a rational ground for those choices themselves.

While almost ubiquitous in academic circles today, this position is not without major difficulties. The first is evident in the language used by Brecht and others who present this argument. Words that indicate what the reader *ought* to do (usually injunctive or evaluative verbs and adjectives) are frequently derived from statements of what *is* the case. For example, Brecht concludes his discussion of arithmetic by asserting: "The logical reasoning in arithmetic is strictly analytic, and it is for this reason that it can *and must be accepted* as *absolutely true*" (*Political Theory*, p. 56; italics added).

Reflection indicates that this position is circular.[33] Why should one use analytical logic (rather than Marxian dialectic, cognitive neuroscience, or divine revelation) to discuss the method of finding the truth? This *use* of logic could itself be interpreted in one of two ways: either it is a deduction, from the "fact" of logical proof to the "value" that such methods "ought" to be used in science, or it is a hidden premise or axiom. Since the former would violate the gulf between *Is* and *Ought*, presumably the value of logical reasoning is itself a premise. But, as Brecht puts it, premises in logic "may be grounded on the belief that they reflect existential reality or self-evident truth" or they may be "based on sheer expediency and even on arbitrary choice": the premises are either beliefs about matters of fact or mere convenience and arbitrary choice. Hence, all logic seems to be derived from a value—but the value itself may be derived from the "fact" that only scientific methods can establish "intersubjectively verifiable" knowledge, or truth.

Brecht's position seems impaled on the horns of a dilemma: either an *Ought* is derived from an *Is* or his premises are purely arbitrary.[34] To avoid this dilemma, many scholars have recourse to the conventional standards of their discipline. Each discipline establishes its norms of methodological and theoretical adequacy; while these are ultimately beliefs, it is both convenient and efficient to observe them. This perspective accepts Locke's view that "men should keep their compacts" because "the public requires it" and applies it to the scientific community itself: the *obligation* to adopt the fact–value dichotomy comes from the *fact* that the scientific method is generally accepted.[35]

For others, the reason for preferring scientific methods and extending them to human affairs is their success in expanding human knowledge and power.[36] Paradoxically, one *ought* to accept the fact–value dichotomy because it *is* effective, useful, or powerful in the natural sciences. Again, however, the practice of the scientist, especially the social scientist, seems to belie the argument supposedly used to define the scientific method.

Finally, some contemporary philosophers of science and many nonscientists openly assert that all science is essentially a matter of value, belief, or purely arbitrary premises.[37] In this view, science and religion are fundamentally similar, reflecting the arbitrary choice of one set of premises rather than another. As I have shown in part 2, however, it is impossible to accept these arguments as conclusive.

Even if science were based on an arbitrary choice of values, this would hardly establish a *duty* to accept arithmetical or logical reasoning as absolute truth. This means that a thinker like Brecht has no grounds for preferring modern science to either ancient science or religious faith, neither of which begins from a sharp dichotomy between analytic and synthetic logic.[38]

The radical distinction between fact and value is thus more problematic than it first appears. The desirability of science is either an ungrounded value or derived from the fact that modern physical science is a success.[39] Either way, the premises of contemporary science and their extension to human affairs need reexamination at a time when the continued beneficence of science and technology is in question.

That the scientific method has been relatively successful over the past three hundred years is not an adequate theoretical justification, especially now that the fruits of the success are open to question. Perhaps it is not an accident that the argument against deriving values (*Ought*) from fact

(*Is*) often rests on mere custom or convention. Many contemporaries take it for granted that the argument against the natural fallacy was established by Hume, but in fact this is not the case.[40]

Hume and the Return to Naturalism

For most contemporaries, Hume's attack on the certainties of reason provides the foundation for denying that existing things (facts) are related to what is proper or moral for humans (values). Curiously enough, however, Hume's *Treatise of Human Nature* does not divorce natural science from moral judgments. Although often cited by writers like Brecht as a supporter of a rigorously value-free approach to science, Hume—along with Rousseau, Hegel, and Marx[41]—is one of the modern thinkers who directly challenge the presumed gulf between fact and value.

Although Hume rejects "eternal rational measures of right and wrong" (*Treatise*, 3.1.2; p. 425), his grounds are not the naturalistic fallacy of linking science and values. On the contrary, Hume's *Treatise* uses a scientific understanding of the passions to show that virtues and vices are grounded in natural "sentiments" or passions.[42] Hume derives what we ought to do (values, the virtues) from observable feelings that can be studied by modern science (facts).

In answering the question of the origin of such feelings, Hume focuses on the ambiguity of the term *nature*. He distinguishes three meanings of the word: opposed to miraculous, opposed to rare and unusual, and opposed to artificial.[43] When turning to the specific virtue of justice, Hume argues that—unlike benevolence and the virtues related to it—justice is one of the "virtues that produce pleasure and approbation by means of an artifice or contrivance, which arises from the circumstances and necessity of mankind" (*Treatise*, 3.2.1; p. 430).[44] Little wonder that Hume is cited as a leading source of the view that science is and must be divorced from values.

Hume's argument at this point is, however, not consistent with the views of Brecht and contemporary positivists. For Hume, justice is "artificial" in the sense that an action needs to be attributed to a moral motive before it can be called virtue. Hume then demonstrates that neither "public benevolence" ("a regard to the interests of mankind") nor "private benevolence" ("a regard to the interests of the party concerned") can be "the original motive to justice" (*Treatise*, 3.2.1; p. 434).[45] The "natural inclinations" of humans lead them to prefer to serve their own self-interest or that of close relatives rather than to attend to the public good. "A man naturally loves his children better than his nephews, his nephews better than his cousins, his cousins better than strangers, where everything else is equal" (*Treatise*, 3.2.1; p. 436). Justice is an artificial virtue based on reason and social experience, albeit a virtue that entails a redirection of the natural virtue of sympathy (*Treatise*, 3.2.3–6; pp. 436–74).[46]

Hume is at pains, however, to emphasize that in the meaning of the word *natural* as "common" rather than "rare," the sense of justice is natural. Indeed, he adds that it is not improper to describe "rules of justice" as "Laws of Nature" in the sense of being common or "inseparable from the species." Hume's position resembles recent naturalistic definitions of the sense of justice based on neo-Darwinian evolutionary theory, not the fact–value dichotomy often used to attack naturalism.[47]

Recently, a number of scholars have emphasized Hume's naturalism.[48] As they point out, Hume—like Aristotle and others in the naturalist tradition—based morality on the wants, desires, or needs of human nature: the *facts* about human sentiments have implications for human *values*.

Hume, of course, recognizes the danger of the naturalistic fallacy: since justice is natural in the probabilistic sense of "common" (as opposed to "rare"), those with limited experience can easily confuse a parochial custom or belief with that which is natural to all humans. But, as McShea notes, Humean naturalists can avoid the naturalistic fallacy if they "limit themselves to the assertion that for a particular intelligent species certain feelings are predictably aroused by certain facts and that the experience of such feelings is the only basis on which we can make evaluative judgments."[49]

A Humean naturalism is not only consistent with the findings of contemporary biology but points to an Aristotelian or ancient scientific perspective as superior to the scientific value relativism described in this chapter. Although Hume (like Aristotle) recognizes the distinction between nature and nurture (or in his terms, nature and artifice), Hume's theory resembles contemporary biology by focusing on the complex interaction between competitive and cooperative behaviors that "necessarily" combine nature and convention.[50]

When reassessed with care, therefore, the premises of the presumed gulf between fact and value are open to question. Neither Locke nor Hume, not to mention more recent thinkers, like Brecht, seems to have definitively established scientific value relativism; quite the contrary, for the substance of recent scientific discoveries points toward a return to a new form of naturalism.

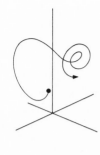

8. From the Naturalistic Fallacy to Naturalism: Neuroscience, Behavioral Ecology, and Mathematical Logic

Is Ancient Science an Alternative to Modern Scientific Relativism?

The notion of a value-free science, so central to the scientific project since the Renaissance, is now in question. In place of logical debates concerning the naturalistic fallacy, it is time to consider in detail whether contemporary scientific evidence supports an approach to science and values akin to that of the ancient Greek and Roman thinkers.

It is important to note what is *not* at issue. At one level, the restriction of scientific discourse to "facts" concerns the difference between science and religion—or rather, the difference between science and those forms of religious doctrine that hold that all events are divinely ordained and hence "ought" to happen.[1] The questions raised here about modern science can hardly be answered in this regard, since to do so would require a resolution of the tension between faith and reason.

This caveat does not apply to the scientific perspective of ancients who did not believe that all human events are divinely willed or predestined by the one God who created the world. For thinkers like Xenophon, Plato, and Aristotle, knowledge explaining the *differences* between one person and another or between one society and another leads to standards of preference. Reason could discover the foundation of right and wrong precisely because humans disagree so profoundly about them.

Do the latest scientific findings favor their ancient naturalism, according to which moral values are derived from knowledge of nature and human nature? To answer this question, the premises of scientific value relativism need to be tested against current work in the fields of neuroscience, behavioral ecology, and logic. In each area, I will summarize the theories underlying contemporary relativism, present evidence challenging currently accepted views, and show the implications of newly established facts for judgments of moral values.

Epistemology: Perception, Cognition, and Judgment

The Theory Underlying Scientific Value Relativism.   Locke and the enlightenment thinkers who followed him denied that there are natural standards of morality and justice in part because without "innate" ideas, there seems to be no means by which reason could discover natural grounds for right and wrong.[2] Individual psychology—the theory of how a human learns about the world—is thus central to understanding human nature and morality.[3] Divine revelation, of course, would remain as an alternative. But insofar as science avoids judgments of "ultimate" or "absolute" value associated with religion, the gap between faith and reason provided a space in which the value-free scientific method could develop without inhibition.

In the *Essay Concerning Human Understanding*, Locke describes "sensation" and "reflection" as the two "fountains of knowledge, from whence all the Ideas we have, or can naturally have" come. He describes sensation and the senses in themselves as passive: they record "qualities" of "external objects," not those objects themselves.[4] We have no direct access by sense perception to what Kant would call the "thing-in-itself" but only can sense external stimuli or "phenomena."

To this passive capacity is, according to Locke, added a second, active "fountain" of information called "reflection."[5] Locke immediately adds that this capacity could be called "internal sense," but "as I call the other Sensation, so I call this REFLECTION, the ideas it affords being such only as the mind gets by reflecting on its own operations within itself" (*Essay*, 2.1.3; vol. 1, pp. 123–24). This active reflection is the source of the categories or ideas of our conscious perception and reasoning about the world. Since natural sensation is merely the material on which the mind operates,[6] such concepts as color and taste are conventions or customs produced by reflection.

Locke's theory was not an isolated statement. Rather, his tabula rasa formed the basis of modern philosophies as diverse as Kantian ethics, logical positivism, and postmodern deconstructionism. In practice, the belief in environmental determinism, which pervaded the foundation of Anglo-

Saxon pragmatism as well as Soviet Marxism, reflects the Lockean view of the brain. But is the theory correct?

The Facts: Neuroscience, Innate Ideas, and Differences of Perspective. The findings of contemporary neuroscience contradict Locke's theory and therewith challenge the premises of modern relativism. For the first time in human history, we can literally see what happens inside the brain when humans (or nonhuman animals) perceive and think. To be sure, Locke was a physician, who relied on detailed observations of neuroanatomy in developing his theoretical understanding of how humans think and know the world.[7] But the methods available to him were crude compared to the PET or NMI scans, event-potential recording from single neurons, neurochemical assays, and fine-grain neuroanatomical analyses that are transforming our understanding of the human central nervous system. Neuroscientific evidence based on these methods reveals that Locke's account is simply incorrect.

In his *Essay*, Locke apparently uses the word *mind* to refer to the central nervous system (or at least the cerebral cortex), and recent research demonstrates that many, though not all, of the "sensible qualities" Locke cites are perceived and discriminated by specialized and localized structures in the brain which are similar in humans, nonhuman primates, and other mammals more generally.[8] Locke uses, as evidence that the basic "ideas" of such fundamental categories as colors and tastes are conventional, the facts that the infant doesn't have them and that they develop only as the child matures.[9]

This argument is clearly inadequate. Since the central nervous system, like other features of the body and behavior, develops progressively, the nature–nurture distinction cannot be equated with the difference between the infant and the adult without greatly confusing empirical analysis.[10] Were Locke's reasoning correct, not only the capacity for language but controlled defecating, walking, eating, and copulation would not be natural or "innate" characteristics of normal human adults.

The discovery of neural networks (discussed above in chap. 5) challenges Locke's assessment in a number of ways. Sensory input seems to be processed in a *hierarchical* manner such that the first processes, at least in the visual and auditory pathways, are highly abstract cues (e.g., straight versus curved lines or movement at each point in the sensory field); at higher cortical levels, features of the stimulus we intuitively associate with reality are reconstituted by distinct cortical structures that discriminate color, texture, shape, movement, and localization in three-dimensional space. These parallel processes are then integrated before we become conscious of a perception. Whereas Locke thought that abstractions such as "straight" versus "curved" were higher-order phenomena resulting from "reflection" on the primary facts of sensory experience, these abstractions are innate; what Locke took as the primary "facts" of sensation seem to be the result of the brain's actively integrating multiple sensory processes.

Perhaps the most interesting evidence of the philosophic implications of this research is the discrimination between "bitter" and "sweet," which, since Protagoras, has been cited as evidence that "man is the measure of all things." In fact, specialized cells on the tongue (taste receptors) have evolved to discriminate between bitter and sweet substances: individual neurons have "distinct response profiles" that are sensitive to the kinds of molecules humans call bitter, salty, or sweet.[11] In other sensory pathways as well, many of the cues corresponding to our fundamental ideas about the natural world are discriminated by localized and specialized neuronal structures. The brain is thus not at all a "blank slate" but rather a highly structured and hierarchically organized system for perceiving and interpreting the world.[12] Humans—and, indeed, mammals generally—have what Locke called "innate ideas" about things as well as about other members of the species.

This is not to say that the perceptual systems of the human brain are invariant or automatic. When discriminating tastes, for example, the response pattern of the specialized taste receptor cells can be shifted if the tongue has been exposed to an appropriate chemical. The individual's experience can thus modify the expression of the inherited sensory capacity. In the same way, of course, exposure of the fetus to alcohol, drugs, or other chemical effects of maternal trauma can produce irreversible damage to the developing brain.[13]

From the perspective of modern biology, innate behavioral traits are rarely invariant, since the observed phenotype is always the product of an interaction between inherited potential (or "reaction range") and environment.[14] Even though there may be genetic predispositions to a trait like alcoholism or Alzheimer's disease, the life history of the individual is critical in determining whether the trait will actually be expressed.[15]

After birth, individual experience continues to alter the structure of the central nervous system, particularly because neurons specialized to respond to cues that an individual never encounters in early development may fail to form synaptic connections and atrophy or die. For example, at birth, the human infant can discriminate the phonemic contrasts used in any known human language. By forty-eight hours, however, the newborn shows a preference for the language spoken in the social environment to which it has been exposed, and by puberty the brain may have largely lost its capacity to discriminate sounds never encountered. This would explain why normal children can learn to speak their native language easily, whereas adults usually encounter difficulty in learning to speak new languages with the appropriate accent.[16] The neuronal structure of the brain is the product of an interaction between natural potentiality and individual experiences that are shaped by culture and society but not entirely reducible to them.[17]

Locke's error seems to have been the assumption that innateness or naturalness must imply *identical* responses in *all* humans. This same mistake arises when he turns to the analysis of "reflection." In speaking of propositions like "white is not black," "a square is not a circle," or "bitterness is not sweetness," Locke argues that the experience of "every man" with reference to "all our ideas of colors, sounds, tastes, figures" and the like would need to be explained by a single process in order to claim knowledge is natural.[18] This implausible equation of the natural with universal and invariant, which can also be found in Locke's view of culture or custom as totally divorced from nature, is contradicted by the scientific understanding of the interaction between innate capacity, individual experience, and social context in the development of cognition.

Since Locke explicitly distinguishes "names" from "ideas," he is not reducing all thought about the visible world to linguistic representations of perception. Instead, Locke asserts that experience and reason necessarily require the "actings of our own minds"—the conscious activities of "perception, thinking, doubting, believing, reasoning, knowing, willing," which are subject to education and cultural shaping (*Essay*, 2.1.3; vol. 1, pp. 122–23). Since individuals differ in their responses to the world, Locke argues that the intuitive "propositions" all humans seem to accept, as well as the abstract "ideas" on which they rest, are the product of active or conscious "reflection," and hence are conventional rather than truly natural.

The interaction between nature and culture, which behavioral ecology reveals at the level of social norms or morality, is paralleled on the level of human perception and thought by the interaction between the brain's innate potentiality and individual experience, the latter itself shaped by both cultural and personal factors.[19] Moreover, Locke seemingly equates learning and education with conscious activity, whereas we now know that much human learning is the product of nonconscious or preconscious information processing.[20]

The unity of an individual's thoughts reflects a complex integration of many factors, some inherited and others acquired, some conscious and others automatic. As has been shown experimentally, much of this integrative process occurs without our being conscious of it and hence cannot be attributed to what Locke calls active or conscious "reflection."[21] The gulf between nature and nurture, on which moderns following Locke elaborated the fact–value dichotomy, has dissolved in the light of contemporary scientific research.

Not only are there innate ideas, but—as was shown in chapter 5—some of these innate ideas directly concern social behavior, forming the basis of our judgment of the feelings, traits, and leadership of others. And since the existence of this capacity makes it rational for selfish or benefit-maximizing individuals to cooperate in otherwise competitive situations,[22] the existence of such innate principles of social behavior in primates generally and humans more specifically is directly relevant to the modern philosophic tradition. The theory of the brain underlying Lockean thought, echoed with modification by Hume and assumed without extensive consideration by most contemporary social scientists, seems to have collapsed.

The Implications for Values: Individual Differences and Social Knowledge. The facts of contemporary neuroscience have direct implications for our judgments and our standards of right and wrong. According to the tabula rasa model of the brain derived from Lockean theory, all individuals exposed to the same situations should perceive the same things. Indeed, this is the central premise on which the methods of modern science rest.

In fact, differences in perception are natural. To cite an obvious difference, the world perceived by a blind or a deaf person will inevitably differ in important respects from that of one with sight and hearing. To cite a less obvious consequence, dyslexics perceive different things when looking at the written page as a result of deficits in the neural structures of language processing that are not due to conditioning or individual learning.[23]

These differences have crucial implications for the way we educate our children and judge each other. Our schools are dominated by intelligence testing, as if there were a single dimension on which human capacities could meaningfully be judged. As Howard Gardner demonstrated, there are at least seven different types of intelligence, each of which has distinct neurological if not genetic foundations. Musical ability, for example, differs from mathematical or verbal ability.[24]

We all know intuitively that this is so. I cannot carry a tune; it would be pernicious to attribute differences between my singing and that of someone with perfect pitch to a lack of effort. Yet our schools often resist applying the same understanding to differences in mathematical and linguistic abilities which also have been shown to rest on neurological substrates. The notion that all citizens are equal before the laws does not mean that all individuals are naturally equal in their potential.

In place of the naïve egalitarianism that has dominated our society, a deeper and more humane respect for individual differences is needed. There has been, for example, a strong movement against the use of sign language by the deaf on the ground that its use would cause them to be treated as second-class citizens. Many of the deaf themselves assert that their condition "is not a disability, but rather a different way of being."[25] And from this perspective, there may be many situations in which it is unethical to treat the deaf and the hearing as if they are entirely the same.

These considerations point to the natural inevitability of differences in perception and judgment. Insofar as personality is related to differences in neurotransmitter function, the perceptual networks and emotional responses of different individuals will inevitably diverge. The methods of modern social science have systematically discounted the natural foundations of such differences, whereas ancient science admitted and emphasized them.[26] By nature, we should *expect*—and *respect*—individual differences.

Beyond the consequences for public policy lies a general principle that is central to both political and ethical thought. If differences in perception and judgment are natural and inevitable, no single individual can claim the truth and no single rule can apply in all cases without prudential modification and individual judgment. In the Kantian tradition, all rational beings should be governed by a single "categorical imperative"; moderns repeatedly seek to solve ethical dilemmas with a single rule applicable to all and understandable by all. For the ancients, opposed perspectives often have an element of truth, and moral judgment involves habits of prudence and moderation that are not universal. Tolerance of others with different views is a moral virtue.

In politics, it follows that controversy and conflict are inevitable. The process of respect for others is necessary, for we can never claim that our own understanding is uniquely privileged. Government under law, political moderation, and the need for political dialogue are naturally preferable to the tyrannical imposition of one individual's will on the entire community. Respect for law and moderation in the use of power are political virtues.

Finally, these considerations have an implication for knowledge itself. Modern philosophers have usually written treatises in which they seek to distill the answers. Among the ancients, one of the central modes of writing was the dialogue. And even when great thinkers wrote treatises, it was to present opposed perspectives in order to distill the elements of congruence underlying theoretical controversy. Humility is an intellectual virtue.

Cultural Relativism: Morality, Virtue, and Obligation

The Theory Underlying Scientific Value Relativism.   For most moderns, the variability of human standards of morality, law, and justice is evidence against the "innate" ideas of duty, whether "imprinted by the hand of God" or derived from "nature" by reason. As Locke puts it, "there is scarce that principle of morality to be named, or rule of virtue to be thought on . . . which is not, somewhere or other, slighted and condemned by the general fashion of whole societies of men."[27] If a "practical rule" is "*anywhere* universally, and with public approbation or allowance, transgressed,"[28] it cannot be obligatory by either divine or natural law.

A similar point is implicit in Hume's rejection of "eternal rational measures of right and wrong" or Brecht's argument that the only "universal and invariant" elements in the sense of justice must be "feelings common to *all* human beings."[29] Cultural relativism—the fact that there are many, mutually contradictory moral and legal rules among humans—is thus said to contradict the belief in a single absolute or universal standard of right and wrong.

The fact of cultural variability leads to an obvious question: is the substance of these rules purely arbitrary or accidental, or are there reasons for the moral and political rules observed in different human societies? For Locke, as for twentieth-century relativists like Brecht, the principal objective seems to be a negative one, namely, to disprove the idea of a single, universal, absolute norm such as the Golden Rule, the Decalogue, or the Law of Love. In this perspective, because cultural norms are entirely conventional, science cannot be the foundation of substantive values.

Not all moderns have been relativists in this sense. Like Aristotle, for example, Hume considers that a science of human culture or morality is as possible as a science of natural things (cf. *Treatise*, Introduction, pp. xii–xvi, with 2.3.10; p. 409).[30] Other moderns have also used science or reason to explain the cultural variability that Brecht uses as evidence in favor of scientific value relativism.[31] These explanations of human culture also provide the basis for standards of right and wrong that are neither universal nor invariant. Could it be that nature establishes changing relationships between the environment and moral or legal norms that explain changes from one time or place to another?[32]

To illustrate, few issues in the United States over the past decade have been as divisive as the claim that abortion is infanticide and therefore immoral. Locke's *Essay Concerning Human Understanding* uses parental care of infants as a central example against the argument that moral rules are innate or natural, citing the widespread existence of infanticide.[33]

A reconsideration of this question from the perspective of contemporary evolutionary theories of social behavior challenges the cultural relativism derived from Locke.

As evidence that moral rules are not innate, Locke gives as an example the rule: "Parents, preserve and cherish your children."[34] He considers parental care of children one of "the most obvious deductions of human reason," consistent with "the natural inclination of the greatest part of men," and hence likely to be "naturally imprinted."[35] Yet Locke says child care cannot be an innate *moral* duty, not only because some people engage in child abuse or neglect, but above all because the obligation to care for offspring is neither "an innate principle which upon *all* occasions excites and directs the actions of *all* men" nor "a truth which *all* men have imprinted on their minds, and which they therefore know and assent to" (*Essay*, 1.2.11; vol. 1, pp. 75–76; italics added).[36] Since it is easy enough for Locke to show that some individuals—and, indeed, often entire societies— have not obeyed the injunction against infanticide and child abuse, his definition of innateness is easily falsified. The problem is that Locke has defeated a straw man.

The Facts: Behavioral Ecology, Social Norms, and Differences of Perspective.   In contemporary biology, research in the field of behavioral ecology shows that the scientific question is different from the one posed by Locke. Darwinian principles lead to variations in the extent to which mammals care for their young. In some cases, net "reproductive success" (especially if measured in terms of "inclusive fitness") may be enhanced by abortion or infanticide. The question, therefore, is whether principles like those observed in other species also influence the development of human cultures and customs.

Among nonhuman primates, the evidence suggests that infanticide and other behaviors regulating birth are adaptive strategies that vary depending on the environment. Among humans, infanticide is typically a mode of population control in response to limited resources.[37] Where culturally established, moreover, human infanticide has been associated with traditions preventing mother–infant bonding precisely because customary infanticide is in tension with the tendency of parents to care for their offspring.[38]

Not only did Locke attack a straw man, but from an evolutionary perspective, variability in customs of infanticide and child care are neither entirely artificial nor arbitrary. On the one hand, humans exhibit individual behaviors (usually called proximate mechanisms) that are conducive to mother–infant bonding and care. These behaviors include a complex set of instinctive responses in both mother and newborn that follow normal childbirth; while these responses of cradling, touching, eye contact, and breast feeding can be disturbed, they are as innate as any inherited behavior pattern among other animals.[39]

At the cultural level, norms or social practices are shaped by the influence of the environment on the tendency to optimize reproductive success (a process of selection usually described as the ultimate cause of observed behavior in animals). Where resources are scarce and unpredictable, humans as well as other species are more likely to adopt strategies of parenting that reduce the investment in each offspring and increase the number of young; conversely, where resources are stable and abundant, after populations grow to the maximum sustainable, there is a reduction in numbers and a relative increase in parental investment per offspring.[40] Cultural variability in the norms permitting infanticide or abortion thus reflects environmental constraints at both the ecological and the social level.

This evolutionary explanation is broadly consistent with the perspective of ancient thinkers, for whom nature was not defined narrowly as invariant and universal. Although *physis* (nature) was indeed opposed to *nomos* (law or convention) by many of the pre-Socratics and Sophists, the term *nature* generally had a different meaning for Plato and Aristotle. The latter, who represent the "old philosophers" attacked by Locke and the modern logical positivists,[41] often treated human nature much as Hume did: that which is natural for humans is *neither* the divinely ordained *nor* the rare, but it *does* presuppose education, thought, and custom. In such a view, nature can be a general tendency (that which occurs usually or for the most part) but also, as in the case of Plato and Aristotle, an end, or preferred state, albeit one that does not occur automatically or with certainty.[42]

For the tradition exemplified in the works of Plato and Aristotle, the relationship between *the one* human nature and *the many* conventional judgments of morality does not reflect a simple dichotomy between nature and convention. Rather, these classical thinkers see nature as the general class within which cultures or conventions arise, and cultures or conventions as the domain within which law emerges. Or to use the excellent formulation of James B. Murphy, this classical tradition rests on the trichotomy of "nature," "custom," and "stipulation" (or law) according to which customary rules are a subset of the natural potentials of humans, and stipulated or legal enactments are a subset of the customs of a community.[43] The oneness of human nature includes moral or legal norms that are approximated by the many diverse instances in which human culture and law respond to the specific conditions of time and place.

The Implications for Values: Cultural Differences and Social Norms.
In the tradition of ancient science, standards of ethics or politics were derived from nature. Does contemporary behavioral ecology have similar
normative implications? It is often charged that this approach is tantamount to the naïve view that "whatever is, is right." Nothing is further
from the truth.

According to the modern view that gave rise to scientific value relativism, teleological views of nature—whether derived from ancient philosophy or from evolutionary biology—cannot be accepted as *scientific* because they entail the assumption of divine mind or will.[44] For contemporary
biologists, the ultimate causes of behavior establish goals or ends, which
are fulfilled in varied ways, depending on circumstance; if so, evaluative
consequences follow without presuming they have a theological sanction.

The vexing issue of abortion illustrates how the naturalistic explanation
of cultural variability can be the foundation of values. On one side have
been those who claim that women have a "right to choose," implying that
no moral standard is more important than the mother's individual decision
based on an assessment of whether or not she chooses to care for the infant.
On the other side have been those for whom the fetus has a "right to life,"
implying that all abortions are infanticide and therefore immoral.

Both of these arguments touch on natural processes, but neither is adequate. Abortion is indeed a mode of population control that, like infanticide, runs counter to the innate responses of mother–infant bonding. But
even if, in a metaphorical sense, one could say that the intentional termination of pregnancy is "unnatural," it does not follow that every unborn
fetus has a right to life; some estimate that as many as 50 percent of preg-
nancies may abort naturally, and all human cultures—not to mention other
species—adopt practices that regulate fecundity as a response to resource
availability.

More specifically, industrial society entails a reduction of the number of
offspring and an increased investment per child. As a result, forcing women
to carry unwanted infants to term entails an institutional requirement that
may lower what biologists call net reproductive success or inclusive fitness.
In less technical terms, legal prohibitions on pregnancy termination may
have the effect of forcibly lowering the economic and social status of some
people compared to others.[45]

On the other side, even if it is in some sense natural to terminate pregnancy when suitable parental investment is impossible, does this lead to a
universal right to choose that has equal moral status in all cases? As a mode
of birth control, conventional abortion is less desirable than the condom

or other methods of contraception (not only for reasons of social cost but because pregnancy always has risks for the mother). Circumstances matter: a married woman's abortion to terminate a fetus with Tay-Sachs disease is surely different from a teenager's routine use of abortion as an alternative to readily available contraception.

In general, naturalistic norms can be stated as ranked preferences whose application depends on circumstances. In regulating human births, because parental bonding is under the control of innate releasing mechanisms, it is natural to seek to minimize investment in offspring whose life is to be terminated. It follows that *abstinence is preferable to contraception, contraception is preferable to abortion, abortion is preferable to infanticide, infanticide is preferable to murder, and murder is preferable to war or plague.*[46]

The danger of conventional ethical criteria lies in their claims of universal applicability as matters of "right." In other species, as well as in most human cultures, the unborn fetus clearly has no right to life—but the mother alone also lacks an unlimited right to choose. These terms convey the claims to resolve vexing moral issues in all cases, yet the consequences can often be unexpected and undesired.

On the issue of abortion, the problematic status of both the right to life and the right to choose is demonstrated by the likelihood that medical technology will make it possible to manufacture human beings to specification. Techniques of in vitro fertilization are well established. Intensive neonatal care is extending viability to earlier and earlier stages of fetal development. And discoveries of human genetics are rapidly pointing to new modes of controlling outcomes.

Ironically, both the right to choose and the right to life make it impossible to reject experiments in biotechnology as unethical. If there is an unlimited right to choose, women would be justified in preferring in vitro gestation (surely justified by the risks of natural pregnancy); if there is an unlimited right to life, once the experiment is begun, the fetus must be brought to term. At the extreme, we would be without guidelines should someone use genetic engineering to design a child who will be a ten-foot-tall basketball player or an energetic but docile laborer.

Futuristic? Alas, not at all. A new technique for the genetic screening of a newly fertilized embryo in vitro has already produced decisions of this kind. For example, a couple has used the technique to screen for cystic fibrosis by fertilizing six embryos in vitro and testing them to choose which one to transfer to the mother for gestation. The outcome: "Two [embryos] had two copies of the cystic fibrosis gene, two had no copies of the gene, one had one copy of the gene, and in one the genetic analysis was unsuccessful. The couple had two embryos transferred [i.e., implanted]—one with a single copy of the gene and one with no copy of the gene. The woman gave birth to a baby girl with no copy of the gene."[47] Such abstract concepts as the right to life or the right to choose are almost useless in deciding how such a technology ought to be used.

The human species confronts a population problem of global proportions. Without a naturalistic perspective, these issues are transformed into competing claims based merely on individual preferences or conventional definitions of right. Worse, presumed rights are cloaked with moralistic rhetoric, without a consideration of the central importance of parental investment as a factor in the lives of all mammals. By converting matters of ethical judgment into universal rights and inflexible rules, both sides of the controversy may unwittingly impose enormous practical and ethical burdens on others and on society at large.

Ultimately, naturalistic norms point to the need for prudence, judgment, and moderation in many areas in which conventional ethical debates lead to intolerance. If circumstances matter, it is not enough to generalize the legal concept of rights into metaphorical solutions of moral dilemmas. But if *only* circumstances matter, then the consequence is a relativism that is both unsettling to most people and contrary to social needs.

Contemporary biological research challenges Locke's argument that both social norms (or what today are called *values*) and the basic ideas and propositions on which perception rests (*facts*) are conventional. Rather, both social norms and individual perceptions are partially innate or natural and partially cultural or individually learned. Knowledge of things and standards of ethics seem to reflect complex interactions between inborn and acquired factors in ways more akin to the Aristotelian understanding of human nature than to Lockean empiricism. But what of the third basic branch of knowledge, Locke's "science of signs" or, in contemporary terms, logic?

Logic: Mathematics and Being

The Theory Underlying Scientific Value Relativism. Since thinkers from Kant to Arnold Brecht base the gulf between *Is* and *Ought* on such logical considerations as the difference between analytic deduction and synthetic reasoning, the foregoing discoveries might be dismissed as technical revisions that do not address the heart of the fact–value dichotomy. Although this argument concerns logic and mathematics, not the empirical sciences, here again recent research demolishes the conventional modern position.

Before addressing the issue, it will be helpful to restate it. As the abortion debate illustrates, moderns tend to formulate moral or political issues in terms of abstract rights. Whether called natural rights, civil rights, or human rights, these claims are universal, replacing prudential judgments of the actor and circumstances with a formal entitlement that has equal applications in all cases.[48]

Such abstract rights thus seem to resolve each specific controversy by a decision that could logically be deduced from general principles. Hence, in the modern tradition, issues of morality are often transformed into the quest for the appropriate rule or verbal formulation. Nowhere is this tendency to substitute logical or verbal analysis for an understanding of the facts more marked than in Kantian ethics.

Unlike Locke and Hume, Kant insists on a priori knowledge. But the synthetic a priori that becomes the basis of the Categorical Imperative converts ethical judgment and morality into an essentially logical operation. To be moral, a rational being (*any* rational being) must act so that the maxim of the will (the noumenon) is a general, universal rule applicable to all such beings. Whereas Locke had denied the universality of moral rules, Kant reasserts them—but transposed from the realm of observed "fact" to that of the "pure" or "free" will. Whereas Hume derived virtue from moral sentiments or feelings, ultimately based in the physical passions humans share with animals, Kant dismissed all such phenomena as impurities that corrupt the free choice without which duty cannot claim the status of a moral act.

The Kantian position thus seems to answer Locke's empirical skepticism about universal standards of moral duty by separating the latter from all questions of fact. This procedure means that moral values must take the form of a rule or principle from which the judgment of each individual case could be deduced. Paradoxically, a similar focus on formulating the correct moral rule on principle, without considering the relevance of objective circumstances or consequences, can be found in British analytic philosophy as well as in logical positivism.[49] But is it possible to reduce ethical judgment to a formula or rule defined in abstraction from human needs and emotions, much like a computer program? To explore this defect of modern thought, consider the Kantian categorical imperative.

The Facts: Logic, Artificial Intelligence, and Moral Agents. Difficulties with Kantian metaphysics arise from the contemporary understanding of reasoning from either an empirical or a purely logical perspective. On the one hand, cognitive neuroscientists have shown that it is literally impossible for a human being to process new information without activating the emotional centers of the brain; on the other, the study of disembodied reason and logic has led to the quest for artificial intelligence—and in the process has demolished the certainties on which Kant himself built his a priori. Each level of the argument deserves attention.

In contrast to Hume's theory of "moral sentiments" (which is broadly consistent with the actual experience of human reasoning), Kant's concept of duty requires that humans engage in acts that we now know to be physically impossible. The Kantian Categorical Imperative requires acts of associative learning and memory, but in humans these processes cannot occur if the cerebral cortex (the site of language, calculation, and mental images, among other functions) is cut off from the limbic system (the structures associated with emotion).[50] How can ethical obligation require a kind of reasoning that is physically impossible?

To avoid absurdity, Kantians tend to take refuge in the purely noumenal realm: the Categorical Imperative is not about how humans actually make moral judgments but about the logic of such judgments. As Kant put it in the title of one of his most famous works, his theory is the prerequisite (*Prolegomena*) to any consistent moral doctrine, not a recipe or a Lockean description of the way the human brain functions. At this level, the Kantian project encounters an even deeper problem.

The Categorical Imperative is a moral rule governing any rational being: can a man-made machine use reason to a degree that makes it what Kant called a moral agent? Popularized by the computer HAL in the film *2001*, the issue is hardly irrelevant in the age of artificial intelligence, neural networks, and supercomputers. Kantians cannot argue that ethical judgment is possible only for a human being, for this considers our species to be a natural kind of rational agent, thereby using a phenomenal category as the foundation of all ethical judgment.[51] It follows that the category of rational agents must in principle be open to nonhuman beings, including computers or other humanly constructed machines. But if so, how could one tell whether a computer thinks in a way that makes it a moral agent subject to the Categorical Imperative?

Whether artificial intelligence qualifies as reason has been hotly debated by philosophers and scientists, many of whom adopt the criterion known as the Turing test (named after the mathematician who first described it). Imagine a qualified observer in one room, communicating by a computer terminal with a second terminal in an adjoining room. If the second terminal is entirely controlled by a computer or other form of artificial intelligence, can the observer at the first terminal distinguish it from messages controlled by a human being? If not, according to the conventional definition of the Turing test, the machine should be considered to "think." Some Kantians, of course, might object that the test is phenomenal (factual) rather than noumenal (purely logical or rational). But if so, one needs to move to a level of pure thinking, which in this case requires a consideration of mathematical theories of automata.

Roger Penrose has recently attempted precisely such an analysis of artificial intelligence by exploring the relationship of pure mathematics, logic, and cosmology.[52] As he shows, the logic of computational devices based entirely on formal algorithms encounters an insuperable difficulty known as the "stopping problem": in theory, any such computer needs a procedure for stopping the computation. The criteria for this action cannot be given by the algorithms that constitute the computational device itself; in existing computers, the program written by the human programmer contains a "stop" command when prior computations have led to a predetermined result.[53]

The human brain solves these problems through pattern matching and emotional responses: as the cortex engages in complex parallel processing of sensory information, the resulting patterns of response are matched to expectations. The role of emotion in stopping the ongoing process of stimulus evaluation probably explains the central role of the hippocampus in memory and learning.[54] In contrast, at least in designs for computers to date, no mechanical device seems to achieve comparable rationality without either introducing arbitrary assumptions programmed by humans or imitating the structures of neural networks.

Elsewhere in mathematics, the very concept of "proof" has come under challenge from demonstrations by computer that can no longer be verified by human reasoning but cannot be discounted on logical grounds. Critics of such computer proofs argue that only proofs verified by human agents can be said to be demonstrable (thereby reducing all rationality to human experience); supporters counter that this amounts to subjecting pure logic and mathematics to the factual—and from a logical perspective, accidental—limits of the human mind.

More generally, it is no longer possible to claim that pure reason or logic can discover a rigorously deductive system that is "absolutely true" (as Brecht claimed). Following Russell and Whitehead, Hilbert sought to find, "for any well-defined area of mathematics, a list of axioms and rules of procedure sufficiently comprehensive that *all* forms of correct mathematical reasoning appropriate to that area would be incorporated"; this project was "effectively destroyed" when Kurt Gödel proved that "any such precise ('formal') mathematical system of axioms and rules of procedure *whatever*, provided that it is broad enough to contain descriptions of simple arithmetical propositions . . . and is free from contradiction, must contain some statements which are neither provable nor disprovable by the means allowed within the system."[55]

Both the Kantian a priori and the apparently distinct (but ultimately related) theories of logical positivism and scientific value relativism are challenged at their very roots by these developments. The dream of discovering a logical rule or mathematical algorithm that in itself solves the problems of either science or morality no longer seems feasible. "Pure reason" cannot even provide an incontestable axiomatic foundation for mathematics, let alone guidelines for practical moral dilemmas. Rationality in logic seems to be of a different order than human thought and action.

The Implications for Values: Logic and Moral Prudence. These limitations of logic are not an abstruse matter concerning only computers and artificial intelligence. At the critical point in his argument for a gulf between fact and value that is based on logic, Brecht introduces the distinction between analytic deductions and synthetic reasoning. His decisive example is the treatment of "mathematical propositions as logical deductions from a small number of basic postulates" so that "the logical reasoning in arithmetic is strictly analytic" and can be described as "absolutely true." At this point, Brecht has a footnote citing the early-twentieth-century project of Whitehead and Russell (among others) to develop a purely deductive mathematics.[56] Today, as indicated by the passage from Penrose cited above, logicians and mathematicians consider this project a definitive failure.[57]

The attempt to develop a convincing logic based on the "pure" a priori has led to a profound paradox: when pushed far enough, any such system requires axioms and assumptions that seem either arbitrary (and hence impossible to accept as "absolutely true") or imitations of nature (and hence empirical hypotheses that return the issue to the realms of biological and cognitive science, empirical adequacy, and therefore the science of natural phenomena). And to resolve these dilemmas, logicians and mathematicians like Penrose find themselves returning to Platonic mathematics—a mathematics of form rather than of algorithm.[58]

In practice, this has two implications for ethical and political discourse. First, information about the nature of things is inseparable from judgments about them. Values adopted without knowledge of the facts are bound to lead to catastrophic errors. In this sense, the critique of "foundationalist" ethics on the grounds that all ethical choices need be unconstrained by the facts is merely a childish wish for unlimited freedom.

Second, judgments about ethical and political matters require prudence. General propositions need to be applied to individual cases. No simple solutions are likely to be universally appropriate. Dialogue among those with different views is necessary. And since not all views will be equally informed or equally reasonable, public opinion polls are often not the proper solution to controversial issues.

Much has been said about the naturalistic fallacy. The question of attributing moral agency to computers and other human contrivances suggests that we need to think about the conventionalist fallacy. If all choices are merely a matter of agreement, is any form of genetic engineering or biochemical control of behavior acceptable as long as some group of humans agrees to it? Is there no difference between a powerful computer and a human being? The deepest danger of scientific value relativism is its inability to control the fruits of the technology spawned by modern science.

Ancient Science and the Return to Naturalism

A generation ago, Leo Strauss articulated with clarity the typically modern dilemma: contemporary science pretends that its discoveries of facts are unrelated to the values by which we live, whereas the traditional values on which our civilization has rested were derived from presumed knowledge about the world. Humans seek some degree of foundation for standards of right and wrong. How then can one combine the need for a teleological understanding of human nature with the rigorously nonteleological view of nature adopted by modern science?[59]

I have suggested that, paradoxically, the answer may depend on the findings of contemporary scientists. The perspective of modern science seems to be based on premises that are not as generally correct or true as has been presumed in the tradition founded by Locke, Hume, and Kant. A return to ancient science seems one way to overcome these limitations of modern logic, modern physics, or modern biology, not to mention the modern social sciences and philosophy. And if so, it would be possible to discover a way out of the typically modern dilemma articulated by Strauss and other contemporary naturalist philosophers. To what extent is a new naturalism, derived from the ancient scientific perspective, consistent with contemporary scientific evidence concerning human behavior and society?

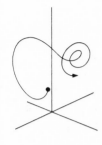

9. Integrating Nature and Nurture: Toward a New Paradigm in the Social Sciences

The Challenge to Conventional Social Science

Prior to the 1920s, the concepts of human nature and evolution had been central to the social sciences.[1] Over the past three-quarters of a century, natural and evolutionary explanations have been excluded in each discipline by the focus on environmental or cultural factors: conditioning in behaviorist psychology; supply and demand in economics; socioeconomic status, cognitive attitudes, and power in political science; popular beliefs, social structures, and conventions in sociology; the "autonomy" of "social facts" and cultural symbols in anthropology.

Historical events have recently called into question the prevailing belief in a profound division between nature and nurture, according to which the study of human affairs should focus on the effects of nurture. Over the past seventy-five years, we have seen the most massive experiment in cultural determinism ever attempted. Soviet Communists used all the resources of behaviorist psychology, social conformism, and political power in the service of Marxist–Leninist theories of human life. Does the failure of this experiment tell us something about the adequacy of the theoretical assumptions on which it rested?

The need to adopt a naturalist perspective on human social life is reinforced by the concepts, theories, and paradigms in a wide variety of disciplines.[2] Most social scientists presume that the natural sciences are relevant only to questions of methodology; substantive research in the biological (not to mention physical) sciences is rarely seen as germane to the study of human behavior. Throughout this book, I have challenged this opinion, introducing evidence that a new naturalism is emerging in the social sciences. To summarize this argument, it will be useful to review recent discoveries and theories in physics, biology, and neuroscience that contradict the notion that humans are somehow apart from nature. What does science now teach about the ways to fulfill the ancient injunction to "know thyself"?

Physics and Scientific Method: Quantum Mechanics, Relativity, and Chaos Theory

Although most secondary-school courses still focus on the classical or Newtonian worldview, contemporary physics has radically transformed the understanding of matter, energy, motion, space, and time. Einstein's theories of general and special relativity destroyed the assumption of a privileged point of observation from which all "objective" measurements can be made; quantum mechanics (particularly in Bohr's formulation) further undermined traditional opinions by indicating that sometimes the fact of observation changes the things being observed. The notion that light can be viewed as either a wave or as a particle—and the nonreducibility of one of these conceptualizations to the other—was but one of the radically unsettling aspects of what came to be known as Heisenberg's uncertainty principle. These changes went beyond a shift from determinism to probability, challenging the very concept of causation at its core.

More recent developments have questioned the conventional understanding of nature even more deeply. Under the label "chaos theory," a new mathematics of nonlinear dynamical systems has emerged as a powerful organizing principle in many areas.[3] Even when such chaotic or nonlinear dynamic systems are determinist, one cannot predict the future of a system from its present state: a set of variables evolving from the same point of origin under the influence of the same parameters can give rise to different and sometimes radically divergent outcomes.

Chaos changes our understanding of determinism, causality, and the world itself. In nonlinear systems, minor perturbations may have enormous and irreversible effects, sometimes at considerable distances in space or time. Shapes or forms—of which the most notable are the beautiful patterns of the Mandelbrot or Julia sets—have unique properties, not the least fascinating of which is self-similarity (the reappearance of a large-scale pattern on a different dimensional scale). Since many phenomena apparently represent fractal dimensions, the three dimensionality of human spatial perception (derived from the brain's processing of binocular visual input) provide only a partial view of nature.[4]

These developments should influence the study of human societies for at least three reasons. First, the scientific methods typically used to study social behavior assume that causal relationships, while probabilistic, are linear rather than chaotic (in the mathematical sense of these terms).[5] In chaos theory, to put it technically, nonlinearity means that there is *no single line* that can connect the data points after a bifurcation or point of diverging outcomes. Causality thus no longer implies predictability, as in Robert May's famous graph of population dynamics (Fig. 3.1). Even more puzzling from the perspective of statistical inference, within the domain of chaotic variability are zones of self-similarity in which the pattern of bifurcation is replicated on a smaller scale. These formal mathematical properties should not be confused with classical statistics; they point to novel modes of interpreting complex and temporally unfolding phenomena.

Second, the introduction of scientific methodologies to the study of human behavior has long been dominated by logical positivism.[6] Although such developments as relativity, quantum mechanics, and chaos theory have led to the general abandonment of this position in the philosophy of science, its methodological principles have persisted within the social sciences. Many social scientists claim, of course, that their work is rigorously scientific. Only a few, however, have stated explicitly that they base their understanding on quantum mechanics and relativity theory. And these scholars, who are presumably in a reasonable position to judge, claim that the social sciences as a whole have ignored contemporary physics.[7]

These observations lead to the most perplexing implication of new research in mathematics and physics. For some (though not all) of those who have reflected deeply on these matters, perceptible space–time is a *reflection* of deeper mathematical relationships. Instead of viewing a mathematical equation as the approximation or "best fit" to describe a material system, one sees the visible material system as an approximation or reflection of the mathematical forms. Hence such leading physicists as Heisenberg, Hawking, and Penrose have concluded that the physics of our time leads us back to Plato and the Platonic forms.[8]

Modern social science has tended to be materialist, adopting either Bentham's critique of metaphysics or Marx's critique of Hegel. Consideration of ideas as prior to material things has generally been viewed as contrary to a scientific study of human behavior. Yet the notion that information has a reality independent of—and in a profound sense constitutive of—material or perceptible systems is characteristic of theories in thermodynamics and cosmology (as well as in modern genetics).

It can properly be argued that developments in physics alone should not determine the appropriate paradigm for the social sciences. The main departures from the laws of classical Newtonian physics are either extremely small, subatomic phenomena or extremely large and cosmic in scale; living beings, on the other hand, are in an intermediate range where neither quantum mechanics nor relativity theory seem directly relevant.[9] What features of contemporary biology challenge the conventional understanding of human behavior?

Biology: Genetics, Evolution, and Behavioral Ecology

Three major features of the life sciences question the prevailing methodological assumptions in the social sciences. First, at the individual level, studies of mammalian behavior contradict the notion that environmental, social, or individually learned variables can be sharply distinguished from biological relationships; second, at the level of groups and societies, a natural science of social behavior is emerging that identifies hitherto unnoticed variables and explains how environmental selection influences both individual behavioral strategies and social structures; finally, at the species level, evolutionary biology focuses on the causes of irreversible change in complex systems, revealing "punctuational" or discontinuous sequences as well as periods of gradual adaptation or modification. All of these phenomena can be understood without abandoning the statistical methods and the theories of classical physics that most social scientists view as their model of the scientific method.

Individual Variability in Behavior. For most social scientists, the operation of cultural variables is radically distinct from natural processes. For biologists, in contrast, any organism is the product of an interaction between its environment and unique life history on the one hand and its genetic endowment on the other. The concept of a reaction range or norm of reaction, generally accepted by neo-Darwinian theorists, represents this *interaction* between nature and nurture by a linear graph of the effect of different environmental constraints on the expression of innate traits or potentials. Of particular importance is the way environmental influences on development modify the physical structure of the human brain.[10]

One can demonstrate the interdependence of natural and cultural factors in many other ways. For a number of traits, observed behavior is influenced by both inheritance and individual life history. At a population level, statistical estimates of these effects of genes and of environments are possible. Using the statistical method called analysis of variance, performances like the speed of thoroughbred race horses reflect the independent effects of inheritance (which characteristically explains between 25 percent and 40 percent of the variation in observed attributes) and of individual life experience or training (which usually accounts for a similar proportion of the variance). Although it may be embarrassing to some, these measures of heritability also apply to many human traits, such as personality and the speed of processing nonverbal information.[11] But populational statistics do not permit us to conclude anything about individuals. To explain the running speed of a single horse or the personality of an individual human, one typically cannot separate "nature" and "nurture" as totally distinct causes.[12]

Contemporary research shows that primate behavior is more complex than was thought a decade ago. Even among chimpanzees, groups exhibit "cultural" variations and individuals differ greatly in "personality." Neurotransmitters associated with behavioral traits in both humans and primates have baseline levels that differ from one individual to another: while there are genetically inherited differences in neurotransmitter activity, these chemicals are also influenced by social status and behavior. Hence both heredity and environment interact in producing the amount of a single neurotransmitter like serotonin circulating in a single person.

The controversial issue of gender differences shows how biological and cultural factors are entwined in ways that are often analytically separable only at an aggregate level. There is now overwhelming evidence, among nonhuman primates as well as in all known human societies, that males are more likely to engage in aggressive behavior than are females. Could gender be *entirely* cultural or socially constructed if—in chimpanzees— males initiate aggressive behavior almost three times as often as females do? Yet it would also be folly to ignore the cultural and social differences in male and female roles, whose variation from one society to another is equally well established.[13]

Social Variability in Behavior.   Contemporary biology also shows how the interaction of biological and cultural factors functions at the social level. Here, paradoxically, what appears to be a form of "methodological individualism" lies at the root of a new approach to societies among all species including humans. Often described as sociobiology, this discipline took shape as a result of William D. Hamilton's path-breaking development of the concept of inclusive fitness. Reproductive success is now defined as the transmission of genes identical by descent (rather than by the number of an individual's living offspring). When combined with the neo-Darwinian concepts of natural selection and adaptation, this approach has given rise to the new field of behavioral ecology, which explains social behavior in terms of cost–benefit constraints shaped by the physical and social environment.[14]

In some ways, behavioral ecology differs little from materialist anthropology or other conventional social sciences. It is now widely realized, for instance, that the cost–benefit considerations of evolutionary biology are similar to those of classic economic theory.[15] Here, however, it must be recalled that a new paradigm often introduces new variables and explains hitherto unexplained or unnoticed phenomena. Evidence is accumulating that behavioral ecology will do precisely this. I cite but one example.

In contemporary political and social science, the economic variable most often used to explain behavior is the absolute amount of income or material resources available to individuals or groups. In behavioral ecology, the *flow* or *reliability* of resources and their *patchy* or *even distribution* in space appear to be more important than amounts in an absolute sense. And in assessing the costs and benefits of alternative behaviors, an evolutionary perspective focuses on the primary role often played by *relative* income— and hence on the importance of social competitors as a distinct variable in social life.[16]

As a result, behavioral ecology points to two factors not typically emphasized by conventional social scientists: the extent to which the physical environment influences the flow and chunking of resources and the extent to which the social environment produces differentiation or stratification in access to resources. As Dickemann, Posner, and others have shown, these two variables can be used to explain the differences in mating systems from one culture to another. Hence, cultural practices like widow suicide, bride-wealth, monastic celibacy, and the sexual double standard—long treated by cultural anthropologists as "social facts" irreducible to any scientific explanation—have become comprehensible as adaptive responses to social and ecological circumstances.[17]

Selection among variants—a concept derived from neo-Darwinian evolutionary theory—can be as relevant to the cultural practices of individuals among human populations as it is to the physical structures of individuals in animal populations. The future will determine whether such evolutionary models are widely adopted, and the outcome will probably depend more on empirical adequacy than on abstract considerations of methodology. Even within science, therefore, changes in theory or paradigm can be seen as a selective process.[18]

Evolution and History. The belief that the definition of each species is entirely conventional or cultural has turned out to be incorrect.[19] A species' gene pool can be studied as a system subject to natural selection even though the selective process operates primarily on the individual pheno-type.[20] Analysis of long-term evolutionary patterns reveals the importance of sudden or "punctuational" change as well as gradual evolution, suggesting new models for human history as well as for biology.[21]

At this level, biologists have begun to use new techniques of DNA analysis. Complex statistical measures of the genetic differences between individuals permit a reasonable reconstruction of the evolutionary pathways that led to speciation. In this way, for example, we now know that divergence between chimpanzees and humans was more recent than the split between gorilla and the lineage shared by humans and chimps. Similarly, these methods show that the pygmy chimpanzee (bonobo) has greater genetic relatedness to humans (i.e., the species diverged more recently) than the better-known chimpanzee.

The difference between analyses of phenotypic and genotypic systems at the level of entire species is not as abstract as it may seem. On the one hand, it gives rise to the discipline of community ecology; on the other, to population genetics. The tendency of social scientists to dismiss entire disciplines of the life sciences is unlikely to remain persuasive over the next generation. This is especially true because the methods used to study genetic variability in human populations also apply to phonemic contrasts in human languages. Within the human species, the genetic and linguistic divergences are directly parallel.[22] Selective and evolutionary processes work on human cultures much as they do on animal species. Languages and the cultures they embody *evolve*.

Neuroscience: The Modular Brain, Emotion, and Consciousness

All serious theories of human behavior have rested on a model of the human brain. From the three parts of the soul in Plato's *Republic* to the dichotomy of eros and thanatos in Freud's *Civilization and Its Discontents*, an understanding of humanity must include some explanation of our ability to think and communicate as well as of our passions and desires. More has been learned about the human brain in the past decade than in all previous scientific inquiry. Whatever one's disagreements on other levels, social scientists must now begin to integrate these findings into their models.

For three centuries, the study of humans has been dominated by the Lockean model of the tabula rasa, according to which conditioning and experience engrave sensory impressions and associations on a blank slate. This view, which denies the existence of innate ideas and treats the brain as an undifferentiated "black box," is empirically false. Its replacement will require serious reconsideration of theories in all sciences of human behavior.

We now know, first of all, that the central nervous system is a parallel distributed processing system. The "modular brain" is composed of specialized structures with specific functional properties. Moreover, the gross anatomic difference between midbrain, limbic system, and cortex explains the relationship between primitive drives, emotions, and complex cognition and relates this triune structure to the evolutionary development of the mammalian brain. In particular, associative learning and memory seem to be impossible if the cortex—the location of complex information processing, language, and motor coordination—is severed from the centers in the limbic system that control emotion.[23]

These structures are regulated by chemicals (neurotransmitters) that function to adjust the signal-to-noise ratio of specific pathways and areas in the brain. Neurochemistry is thus becoming the key to understanding sexuality, mental illness, crime, alcoholism, and personality. Neurochemistry even influences politics, since serotonin plays a central role in dominance behavior and leadership among humans as well as other primates.[24]

Cognitive neuroscience challenges the conventional view of mind–body dualism and human consciousness. It is no longer possible to assume that all cognitive processes are under conscious control; rather, preconscious information processing occurs in parallel as different sensory modalities and feature detectors respond to the environment. Language is a specialized human capacity, based on a distributed set of specialized modules in the cortex. When parallel processing gives rise to discordant perceptions or responses, what Gazzaniga has called the "interpreter module," a structure in the left hemisphere that is independent of language or consciousness, apparently resolves the discrepancies traditionally described as cognitive dissonance.

These findings have immense importance for our understanding of human consciousness, not to mention their implications for epistemology and moral philosophy.[25] Perception of many features of the natural and social environment depends on neuronal ensembles that have an innate capacity to make specialized discriminations. With regard to social behavior, for example, individual neurons in the temporal lobes are programmed to fire only on perceiving a cue like the upward movement of the head, associated with facial displays of anger/threat. Because nonverbal expressions of emotion that communicate meaning are processed by these evolved structures in the brain, such notions as happiness or anger can properly be described as innate ideas.[26]

Consciousness often reflects rationalizations that bring harmony to otherwise discordant responses, many of which are themselves beneath the individual's awareness. Moreover, as evolutionary theorists point out, deception of social competitors typically has selective advantages, but the capacity to express emotion generates a countervailing tendency to "leak" cues of the deception. As a result, the most reliable way to deceive others may be to deceive oneself. Hence, as moralists have long noted, humans are peculiarly susceptible to self-serving and manipulative behavior disguised as disinterested altruism.

Conclusion: From Method to Substance

Most social scientists, when encountering the evolutionary perspective on human affairs, have ignored it or tried to suppress it, using the anonymity of peer review to oppose the publication of scholarly work they neither understand nor respect.[27] My contention is that the social sciences in general, and political science more specifically, need a new paradigm. This claim is based on the wealth of substantive scientific information in recent physics, biology, and neuroscience that contradicts prevailing methods and conceptualizations.[28]

It is hard to know how such decisions can be made on other than scientific grounds, based on the substantive theories and findings of all the disciplines relevant to the phenomena under study. That means, in short, the need to focus on substance rather than method. And, I would insist, this cannot be done without taking the time to learn something concrete about contemporary physics, biology, and neuroscience.

One can, of course, persist in using an obsolete paradigm. It is probably possible to describe the motions of the solar system using Ptolemaic astronomy provided one is willing to use enough epicycles. But at some point, the awkwardness takes its toll. The new perspectives not only explain much that was unintelligible in the past but even make better sense of conventional wisdom. It is hard to believe that this will not occur over the next generation. As an increasing number of scholars are realizing, it is impossible to understand social life and politics without a knowledge of nature and of human nature.

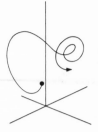 Conclusion: Postmodernism
and the Return to Naturalism

My thesis can be summarized briefly. The conventional understanding of modern science is contradicted on key points by contemporary evolutionary biology, behavioral ecology, neuroscience, mathematical physics, and logic. In contrast, ancient scientific and philosophic conceptions of mathematics, of form, and of human nature often seem consistent with recent scientific results. Ironically, the process of following the methods and theoretical assumptions of natural science leads us back to the perspective of ancient Greek thinkers from Thales to Aristotle—that is, to a position that has been rejected on supposedly scientific grounds since the seventeenth century.

For the general public, the central issue is doubtless the value-free conception of social science. Most Western scholars today take it for granted that a logical gulf between *Is* and *Ought* precludes the establishment of values or moral standards on scientific grounds. The ancients, in contrast, taught that knowledge of nature and human nature points to the proper way a person should live. Properly understood, the principles of naturalism can avoid the naturalistic fallacy and establish the foundations of moral values.

The Limitations of Modern Science

My analysis reveals five principal limitations of the assumptions and methods of modern science.

• First, the *historical relationship* between science and the social context in which it occurs. Modern natural science emerged in a specific cultural and political setting and may not be exportable to cultures unlike the Western societies in which it has flourished (see chap. 2).

• Second, the inadequacy of a *value-free conception of scientific knowledge*. Contemporary research in behavioral ecology, neuroscience, and logic undermines the claim that modern science either rests on or establishes a gulf between *Is* and *Ought*. As the ancients taught, knowledge of nature is at the foundation of human values or goals (see chap. 8).

• Third, the ubiquity of *nonlinearity and chaos* in natural phenomena. New mathematical approaches provide a general model within which the linear regularities of Newtonian physics seem to be a special case. The focus on form and geometry characteristic of ancient science has not been definitely replaced by algebraic algorithms based on the assumption of temporal or spatial linearity (see chap. 3).

• Fourth, the error of the *nature versus nurture dichotomy*. It makes no sense to ask whether many behaviors are the result of *either* nature *or* nurture: genetics, organic development, individual learning, social tradition, and ecology all influence human life, and each factor has both natural and cultural dimensions. Contemporary biology indicates a reasonable way of understanding the teleological view of nature developed in ancient science (see chap. 9).

• Fifth, the impossibility of a definitive *conquest of nature*. There are limits to the human ability to predict and channel natural or social outcomes. The optimistic view of history accepted by moderns in the Baconian tradition can no longer be taken for granted (see chap. 1).[1]

These limitations of the modern approach to scientific knowledge justify a reconsideration of the scientific perspective developed by ancient Greek thinkers.

The question of the proper goals or aims for human activity was central to ancient science and philosophy. It is now imperative to reopen this issue for at least five reasons, which parallel the limitations to the modern scientific approach:

• First, our *historical situation* compels us to a new examination of how scientific discoveries influence human purposes and values. the combination of the human genome project and genetic engineering offer the possibility of using scientific knowledge to reshape the species. Even without such changes in the basis of human nature, developments in neuroscience will make possible hitherto unthinkable manipulation of behavior. And even if we forgo such techniques, modern medicine and the technology of reproduction have already altered human life in fundamental ways, generating deep conflicts over contraception, surrogate parenting, abortion, the cost and distribution of health care, and the effects of prolonged human life. Protection of the environment in the interest of future generations, while easy to proclaim as a goal, is difficult and highly controversial in practice.

• Second, the concept of *value-free science* poses a practical problem. Controversies over the goals or uses of scientific or technological knowledge influence the funding and autonomy of scientific research itself. The social constraints on science are particularly evident whenever the human consequences of a discovery become matters of public policy, for then the specialists are recruited into political controversies. When this happens, the behavior of the scientist, even if qualified verbally, often contradicts the philosopher's image of a gulf between *Is* and *Ought*.[2]

• Third, the *nonlinear or chaotic* character of social processes makes it difficult to predict the consequences of technological innovation as well as natural phenomena from climatic change to the AIDS epidemic. Unexpected negative outcomes undermine the public's willingness to trust scientific predictions. A culture based on the dream of a "conquest of nature" is singularly unprepared to respond to natural catastrophe or political impotence.

• Fourth, the interaction of *nature and nurture* makes possible a deeper scientific understanding of human behavior. This knowledge comes at the cost of complexities that are difficult for most citizens to understand. Conventional social sciences are no longer adequate as the sole background for explaining many cultural and political phenomena.

• Fifth, the *limits of our control over nature* point to a growing realization that the concept of unlimited historical progress is a myth. Humans need to live within nature. Even if we colonize outer space, engineer changes in our own species, or inadvertently destroy life on earth, such radical transformations of the human condition will not be "conquests" or victories.

From a scientific perspective, it is necessary to reconsider the possible contribution of ancient science and the view that values or goals may be derived from nature. But these changes are also desirable from a humanistic point of view.

The Dangers of Postmodernism and Relativism

A naturalist understanding of human life is relevant to the average citizen as well as to the scientist. Postmodernism—the view that all knowledge, including science as well as religion, is culturally conditioned—has spread in our universities, particularly among those who study literature, philosophy, and the arts. Critics of this movement have defended traditional moral standards without effectively answering the claim that all human values are relative. In politics as well as in private life, there is a persistent debate concerning the "old values" and their relationship to new realities.[3]

The extreme relativism of contemporary culture is not limited to a few professors. In one popular advertising slogan, "Just Do It," the "it" is undefined, since each consumer seeks individual goals. Beneath this openness is, however, the hedonism implied by the sexual connotation of "doing it." Our commercial society reveals its value relativism from the debate over multiculturalism in education to the diversity of our cities. The social "openness" of our epoch reflects rapid and sometimes bewildering changes in moral standards.

When I was growing up, contraceptives were illegal in some states and only available to married couples in others; today birth control pills are ubiquitous, and the debate is whether condom machines should be installed in high school lavatories. In our colleges, a generation ago women students had to sign into and out of their dormitories and could not stay in a man's room overnight; now there are coed dorms and the old parietal rules are of only antiquarian interest. In such times of rapid sociocultural change, it is not surprising that traditional values are attacked as representing the interests of some segments of society against either the needs of others or the collective good of the community.

The extreme relativism of many intellectuals needs to be placed in this social context. When accepted standards of right and wrong confront new technological and social realities, change becomes inevitable. Postmodernism can be understood as an attempt to legitimate change by challenging all previously accepted norms as relative to time, place, and self-interest. The difficulty is that, in the long run, humans cannot live without values, and values need to be based on something beyond intuition, whim, and feeling.

The problem is evident in the paradox of humanistic disdain for science in a highly scientific culture. Postmodernists write relativist criticism using personal computers, apparently ignoring the implications of the science and technology that has given rise to the pressure for changes in the way we think. The spread of cultural relativism over the past generation is understandable. As we look to the next century, it is also dangerous.

The humanistic approach to the division between scientific knowledge and human purpose encounters theoretical and empirical difficulties parallel to those confronting the fact–value dichotomy. The postmodern attack on scientific objectivity, derived from a vulgarization of Kant and Nietzsche, takes the form of extreme cultural relativism. For literary critics, philosophers, and many intellectuals, this means that human ideas, beliefs, and values are entirely subjective: no knowledge or interpersonal truth whatever is possible with regard to human affairs.

In this currently fashionable view, even philosophic texts cannot be interpreted with certainty. It is claimed that we can never know the intention of another human being and, as a result, never decode accurately the meaning of another's words (especially if the author lived in other times and places). In place of truth, the postmodern temperament substitutes a personal interpretation or reading that *deconstructs* the text without claiming knowledge about its substance. In place of general principles, the postmoderns speak of *stories* or individual events. In place of a common core of knowledge that an elite needs to understand in order to govern effectively, emphasis is placed on *diversity* and the need to give equal attention to every voice.[4]

This postmodern temperament can be traced to a combination of factors. The "democratic" or egalitarian generosity that characterizes the left or liberal impulse in modern politics inclines many to an openness to novelty. At the same time, however, the commercial emphasis on individual profit that characterizes the right or conservative position encourages the libertarian view that governments or societies should not control private choice. Most important of all, the social and economic abundance and stability of Western industrial societies during the cold war predictably generated extreme individualism and a decline in the deference to traditional moral values.[5]

This explanation of the origins of postmodernism is not a defense of its principles. On the contrary, such a radically antiscientific attitude is tantamount to intellectual suicide at a time when the biological sciences make it possible to manipulate human behavior and genetics for political purposes.

Nihilism may have seemed a harmless and pleasant luxury for the intellectuals of advanced societies in the 1920s, but it is surely folly today. We must look forward to the next century, when those in power—armed with hitherto undreamed of technologies—could shape entire societies to any whim or will. Postmodernists and their critics, who have debated the flirtation with Nazism by Martin Heidegger or Paul de Man, now need to direct their attention to the practical dilemmas that will almost certainly face their children and their students.

The extreme skepticism, relativism, and nihilism of postmodernism is but an exaggerated form of the characteristic self-assurance of the modern scientific project and its presumed conquest of nature. If the ends or purposes of life are beyond discussion, all that matters are the means used to achieve each individual's will: *do your own thing*. But the modern view of science is not the only possible perspective accessible to human knowledge.

It was an act of hubris to imply—as Hobbes, Locke, and so many moderns have—that their perspective could totally supersede both the reasoned dialogue of ancient science and the faith of those who believe in revealed religion. On the contrary, these means of seeking knowledge about the ends of human life provide a necessary complement to modern scientific methods and discoveries. Before summarizing the values derived from this naturalist perspective (see "Naturalism, Excellence, and Values," below), however, a word is in order on both ancient science (which has been the focus of this work) and religion.

The argument that moral values can be derived from knowledge of nature and human nature is meaningful only if we understand the theoretical position represented by naturalism. Ancient science—or, as it is conventionally called, ancient philosophy—has a claim to truth that has long been dismissed on the assumption that history is progress. Although it is difficult for us to recover the perspective of the ancients, the effort is essential if naturalism is to provide a reasonable framework for moral dialogue. How, then, did the ancients relate facts to values (see chaps. 3, 4 ["Ancient Science, Pattern Matching, and Knowledge as Seeing" and "Intuition, the Socratic Question, and the Dialectic"], and chap. 8)?

Ancient Science and the Problem of Values

For the pagan Greeks, what is natural occurs "for the most part" or usually. Both Sophist and Socratic thinkers consider law (*nomos*) as humanly instituted conventions or rules that vary according to time and place. Because human conventions differ in substance from the rules of inanimate nature (*physis*), the Greeks usually do not speak of "laws of nature": on the contrary, the humanly stipulated "rules of the laws" can be contrasted to "rules of nature," not only for a Sophist like Antiphon (for whom society is contrary to nature) but for Aristotle (for whom our species is the "political animal").[6]

Greek thinkers with divergent theories of human society and justice nevertheless agreed that nature is the standard for judging human life. For the Sophists, since society is unnatural, it is "best" for a human to follow the rules of nature—either avoiding social obligation (Antiphon) or manipulating society for individual benefit (Thrasymachus). In the tradition of Plato and Aristotle, this understanding cannot be adequate because humans always live in societies governed by customs and laws (see "The Origins of Society" in chap. 6).

Despite this disagreement, all of the ancients understood nature as neither rigidly determinist nor absolutely universal: accident or external necessity can often cause natural things to depart from their usual or proper condition.[7] For the ancients, this condition or end of a thing—its natural or intrinsic form—is the basis of our values and preferences.[8]

The tradition inaugurated by Socrates and immortalized in the works of Plato, Aristotle, and their schools[9] diverges from other ancients (not to mention many moderns) in viewing speech and the division of labor as evidence that humans are naturally sociable (see chap. 5). Since the natural end of a species is the fulfillment of its potentialities, this Socratic tradition teaches that political life—and such political or moral virtues as justice, moderation, or wisdom—are in this sense "according to nature."

The position of these "old philosophers," as Locke called them, differs from both Christian theology and modern science (see "Locke and the Origins of Value-Free Science" in chap. 7). For pagans who did not believe that a monotheistic God formed the world in six days, the natural virtues are not the intentional plan of a conscious mind acting as omnipotent and omniscient creator of the universe. Rather, the ancients based ethical standards on a rational understanding of "the highest perfection of human nature" and "the dignity of man" (Locke, *Essay*, 1.2.5; vol. 1, p. 69).

Aristotle provides the most explicit statement of the view that natural values or ends are *not* due to an act of mind or "will" imposed on neutral matter. As Aristotle puts it, nature is like "a doctor doctoring himself":[10] the principles of order are immanent in nature, not extrinsic or immaterial. Perhaps for this reason, Aristotle probably provides the deepest view of biology prior to Darwin and his successors (see chap. 5).[11]

The Aristotelian view of human nature points to the exercise of social and political life as an end or perfection of the species' abilities for speech, intentionality, and conscious self-control. Courage, moderation, justice, and wisdom are "by nature" in the sense that humans naturally have the ability to control and direct passion through cognition and reason, even though to do so requires proper education and customs. Goals or values capable of guiding human life can be derived from a teleological but non-theological "naturalism," but they cannot be imposed arbitrarily or reduced to rigid doctrines.[12]

Variation in human customs and morals is one of the strongest arguments *in favor of* a naturalist approach to ethical and legal philosophy.[13] Locke's attack on innate ideas, while perhaps effective as a critique of divinely ordained natural law, does not address the human capacity to reason and live in complex social systems as the defining characteristic of our species. Modern theories cannot provide adequate standards of judging social phenomena because they rest on an imperfect understanding of human nature (see chaps. 5, 7, and 8).

One can best see how the ancient perspective differs from modern science or postmodern philosophy by considering concrete instances. According to Aristotelian naturalism, the art of medicine is the model for the way scientific knowledge responds to practical judgment.[14] Nature provides the standard of health, though illness occurs by necessities extrinsic to the best form of an individual. Since nature is not equated with necessity (as it is for the moderns), one can consider a disease necessary without concluding that it is the nature of the organism to be ill. The facts of health guide the physician's practice when treating the patient's disease; the facts of human nature guide the philosopher's judgment of social norms and rules.

A simple example indicates the superiority of this ancient view to the modern fact–value dichotomy. When a physician advises a patient to have an operation for appendicitis, the patient is unlikely to complain that it is a logical fallacy to derive the value of the operation from the fact of the disease. Although appendicitis happens "in the nature of things" (i.e., as a necessary consequence of physiological or natural events), it is not a condition that characterizes the ends or purposes of humans as distinct from other natural things.

Critics of this example explain the desire or value attributed to health as a "prescriptive premise" (using terms from logic) without which the diagnosis of appendicitis (a fact) would not lead to the preference for the operation (a value). Such a criticism ignores the evolutionary process that led to human desires for a healthy life. While patients in some situations (such as the aged who are terminally ill) do *not* wish to survive, the reasons for such choices are knowable, and these reasons are essential to any prudential and ethical judgment of the case.[15]

The problems of medical ethics that have resulted from the prolongation of human life indicate the difficulty of the modern view of abstract "rights" and principles. Many contemporary ethicists have adopted a contractual view of the relation between physician and patient: treatment should depend on "informed consent," on the assumption that the patient's act is "free," or unconstrained. But choice is always influenced or caused by something, and few of us prefer to be ill.[16]

From an Aristotelian point of view, the fact that some individuals choose to refuse treatment or even to commit suicide need not be an ungrounded act of will; rather, such choices often result from natural processes that influence the cognition and judgment. Health care needs to be understood in terms of the end or goal of health. And health needs to be understood as the normal, natural function of the body and mind. The contractual notion of medical ethics, like the contractual theory of society, is limited by its replacement of reasoned dialogue about ends with the presumption of an unconstrained or free will.[17]

Humans have a single nature as a species, even if individual capacities and choices are widely different from one place or person to another. What is proper to this nature is its best condition or end. For a small number of humans with the capacity for scientific research, this end includes the quest for knowledge concerning natural things, a quest that is difficult if not impossible for others.

Dispute about the ends of human life indicates how important it is to discover what is according to nature under given circumstances. For the ancient Greeks generally, but especially for Aristotle, a functional or teleological view of nature means not only that value judgments need to be based upon the facts but that this process entails reasoned dialogue among reasonable people.

Religious Faith and Reason

Before summarizing naturalistic values that can answer the challenge of relativism, it is necessary to consider one final objection. Is it not the case that controversies over theological dogma make it impossible to reach agreement through "reasoned dialogue among reasoned people"? Religious hatreds like those between Catholic and Protestant in Northern Ireland, Muslim and Jew in the Middle East, or Hindu and Muslim in South Asia seem evidence that definitions of human values based on faith contradict the premises of a new naturalism.

I have argued that there are three fundamentally different perspectives toward knowledge, corresponding to explanations of human morality on the basis of convention and custom (modern science), reason and nature (ancient science), or faith and God's will (revealed religion). In this essay, I have focused on the need to complement modern science with ancient science. Were it appropriate, however, this argument could be extended to the dangers of seeking to impose reason (whether in its modern or its ancient form) on the entire domain of faith. Superficial intellectuals often seem to claim that one of the three principles of meaning should replace the other two. Perhaps the truth is that each is in part irreducible to the others. If so, human excellence now requires a civilization that sustains the tension between reason and faith as well as the tension between the ancients and the moderns.

That the methods of modern science can coexist with both an understanding of ancient science and a faith in God is demonstrated by a number of great thinkers in our tradition (including Pascal, Rousseau, Plato, Ruskin, and Tolstoy). To illustrate how the difference between reason and revelation can be respected without dismissing either ancient or modern science, it is particularly valuable to reflect on the thought of Pascal.

Although I have stressed the continued validity of ancient science and the limitations of modern science, Pascal emphasizes the need to combine both and relate them to religious belief.

Because Pascal was openly religious, one might expect him to favor ancient science over that of the moderns, but this was clearly not his intention. Not only does he seek to combine the excellence of ancient reason (what Pascal calls *l'esprit de finesse*, a difficult term translated as "intuitive thinking") with the characteristic focus of modern science (*l'esprit de géométrique*, or "mathematical thinking"),[18] but he also emphasizes how rational inquiry is useful in turning those with despair toward faith. Mathematical thinking (*l'esprit géométrique*) and intuitive thinking (*l'esprit de finesse*) need to be distinguished from "false ways of thinking" (*les esprits faux*), and the surest route toward that end available to human reason is the combination of the two methods represented by ancient and modern science.[19]

Pascal's position symbolizes a way to balance competing perspectives. Because judgment requires intuitive thinking whereas modern science relies on mathematical thinking, Pascal argues that they should be combined in a way that does not exclude faith. A faith that dismisses all reason as pride can neither avoid the risks of human passion and corruption nor persuade those in despair. A reason that ignores the modern, scientific mode of analysis will always be partial. But today, at least in American intellectual life, the critical danger would seem to come from a different quarter: a reliance on the modern scientific project (Pascal's *esprit géométrique*) as if there were no other foundation of truth, morality, or wisdom.

The division of academic life into hermetically isolated disciplines has permitted extraordinary advances in the knowledge of what Aristotle calls "material" and "efficient" causation and therewith unsurpassed possibilities for technological innovation. It does not, however, have the capacity to establish a permanent "conquest of nature." Quite the contrary, as C. S. Lewis insightfully noted, "what we call Man's power over Nature turns out to be a power exercised by some men over other men with Nature as its instrument."[20] The ultimate form of such tyrannical use of scientific knowledge would, of course, be the accidental or intentional destruction of the human species as we know it. The task for coming generations will be to discover whether the fruits of modern science can be controlled before their use destroys humanity.

Naturalism, Excellence, and Values

What, then, *are* the moral values derived from the new naturalism proposed as a remedy for relativism? Although the answer depends in part on circumstances, unlike rigid doctrines of theological dogma or natural law, our common humanity does point to a core of common principles that can claim to be natural. When considering the normative conclusions from a scientific study of human nature, however, it is important to bear in mind that their application will depend on concrete circumstances.

In general, the standards of right and wrong associated with naturalism are the common elements found in the teachings of all major world religions and ethical doctrines. From a religious perspective, C. S. Lewis describes them as "the way" (the Tao); in a more secular vein, James Q. Wilson calls them "character" in the sense of moral responsibility and rectitude.[21] The classical Greek and Roman thinkers spoke of courage, moderation, wisdom, and justice; Christianity added faith, hope, and charity. "Do unto others as you would have them do unto you."

The core of naturalistic values can be stated more concretely. The traits of human excellence in this generic sense are those one would choose for a friend. No one would wish that a true friend lie, cheat, or fail to help at a moment of need. In friendship, we all seek sharing and caring that goes beyond the text of the contract or the letter of the law.[22] Even a nihilist or a tyrant does not seek out a scoundrel for his best friend.[23] Nor is friendship the same as contractual agreement to engage in a mutually beneficial exchange; the latter is typical of everyday purchases in the market economy, whereas a true friend is interested in one's welfare even at a net cost to his own time and effort.[24]

I can be even more precise, at least with regard to moral values (Aristotle's "practical virtues").[25] At the outset of this chapter, I noted five limitations of modern or value-free science. Naturalistic values are associated with an understanding of each of these five factual conclusions:

First, the *historical relationship* between science and the social context in which it is discovered is evidence of the differences between one human society and another. This fact leads to the obligation to *respect the existence of social and moral norms of cultures unlike our own*. If circumstances matter in judging human conduct, we need to hesitate before assuming that our interpretation of the common core of human nature entails criticizing the way of life of another civilization. This is not, however, "anything goes": cultures that assert the right to annihilate other peoples, races, or religions thereby implicitly claim absolute truth for their own principles and thus violate this norm. Respect for diversity, proclaimed as a right by postmodernists, is far more soundly founded in naturalism than in cultural relativism.[26]

Second, the inadequacy of a *value-free conception of scientific knowledge* is evidence that humans have a deep psychological and social need for

values. Relativism leads to a self-defeating circularity, since we cannot live
without expectations concerning the behavior of ourselves and of others.
For most people, life without moral or ethical principles is not possible.
The sense of justice—described by biologists as "moralistic aggression"—
is, as a matter of fact, natural to the human brain.[27] The pretense that one
can avoid the need to make moral choices is in itself a moral choice. More
important, the fact is that *each of us is responsible for our moral choices*
(if only because others will inevitably behave as if we were responsible).

Third, the ubiquity of *nonlinearity and chaos* in natural phenomena
reminds us that many of our actions are certain to have unintended and
unexpected results. We are by nature responsible for what we do, if only
because others naturally hold us responsible. But because we are not totally
able to predict the consequences of our own actions, we are obligated to
consider the possibility that the deeds of others were in part unintended.
*Toleration and forgiveness of the errors of others is the counterpart of a
willingness to take responsibility for our own actions*; both are required
by the Golden Rule.

Fourth, the error of the *nature versus nurture dichotomy* means that
education is necessary for the development and perfection of natural potentiality. Excellence (in Greek, the word was *arete*) is virtue. But while each
of us is responsible to develop our natural potential to the fullest, this
potential differs considerably from one person to another: one is a fine
craftsman, another an extraordinary mathematician, the third a gifted per-
former. We need to *tolerate differences in ability while always demanding
more of ourselves than others are justified in asking*.[28]

Fifth, the impossibility of a definitive *conquest of nature* means that inevitably some social conflicts will be impossible to resolve by increasing
the resources or power available to all. The facts of human existence ensure that there will be competing claims that someone has to adjudicate.
This leads to the value of *natural justice*: we are obligated, insofar as is
humanly possible, to seek the just or fair resolution according to nature.
Plato saw justice as occurring in the individual when each part of the soul
(or personality) fulfills its natural function; he defines justice in society as
the condition in which each social class plays its proper role. In place of
limitless claims on others in the name of natural or civic rights, morality
asks of us self-imposed obligations to achieve justice.

These standards of right and wrong are not rigid. Nor are they plati-
tudes. As I write these words, the news gives us ample evidence, from
Croatia or Bosnia to Somalia, Haiti, and Kurdistan, of the ease with which
passions overcome the obligations of natural justice. A nihilist or relativist
cannot condemn self-interested leaders who use racial, ethnic, or religious

particularism to enhance their own power at the price of others' lives: it hardly makes sense to condemn genocide as uncivilized if civilization itself is without value. The fact of scientific knowledge, itself dependent on civilized society, leads to an obligation to achieve that which is just "according to nature."

An essential caveat: The course of action called for by naturalism in any circumstance may (and often will) be subject to controversy. The same situation will lead to different assessments based on intuition, hypothesis testing, or pattern matching (and the correlative perspectives of religious faith, modern science, and ancient science). The ubiquity of moral and political controversy does not contradict naturalism; rather, it demonstrates naturalism's soundness. But prudence is needed to balance these different assessments, and some individuals will be more prudent than others. Respect for the dignity of all other human beings does not mean that all people are equal in ability (and especially in the ability to make moral judgments of difficult cases).

No simple rules provide an unproblematic, universally correct foundation to moral choice or political values. Neither Kant's Categorical Imperative nor Rawl's "veil of ignorance" can admit of the subtlety of variation and exception required to apply naturalism in every human condition. If none of us can achieve perfect justice, all of us can move as close toward this ideal as possible.

Plato, in concluding the *Republic*, has Socrates state the point as well as it has ever been stated:

. . . each of us must, to the neglect of other studies, above all see to it that he is a seeker and student of that study by which he might be able to learn and find out who will give him the capacity and the knowledge to distinguish the good and the bad life, and so everywhere and always choose the better from among those that are possible. He will take into account all the things we have just mentioned and how in combination and separately they affect the virtue of a life. Thus he may know the effects, bad and good, of beauty mixed with poverty or wealth and accompanied by this or that habit of soul; and the effects of any particular mixture with one another of good and bad birth, private station and ruling office, strength and weakness, facility and difficulty in learning, and all such things that are connected with a soul by nature or are acquired. From all this he will be able to draw a conclusion and choose—in looking off toward the nature of the soul—between the worse and the better life, calling worse the one that leads it to becoming more unjust, and better the one that leads it to becoming juster. He will let everything else go.[29]

The apparent redundancies in this passage are clearly intended.

Plato wishes to convey the complexity of the facts that need to be determined in order to act justly. He leaves open the issue of exactly who it is "who will give him [each of us] the capacity and the knowledge to

distinguish the good from the bad life": perhaps some humans lack this capacity, whereas others have it. But above all, he emphasizes the importance
of circumstance: naturalism asks us "everywhere and always *to choose the
better from among those that are possible*" (italics added).

The reconciliation of contemporary science and traditional philosophy
entails understanding the limits of nature and of human nature. Although
it is only within these limits of possibility that we have an obligation to live
according to virtue, science can indeed lead us beyond relativism toward
the discovery of what the ancients called natural justice.

Notes

These notes should help resolve the difference between the needs of the general reader and those of the scholar. They will provide not only references but added explanations of complex issues as well as textual citations from the philosophers whose arguments I summarize. Specialists, philosophers, and critics should therefore consider the notes, while the reader anxious only to follow the main line of argument need not consult them unless seeking specific references. For the model on which this usage is based, see Jean-Jacques Rousseau, *Discourse on the Origin of Inequality*, "Notice on the Notes," in *Collected Writings of Rousseau*, ed. Roger D. Masters and Christopher Kelly (Hanover, N.H.: University Press of New England, 1993), Vol. 3, p. 16.

1. The Crisis of Modern Science and Society

1. It would be easy but pedantic to document each of these observations in detail. There is much evidence of poor performance in the sciences by American students, but most proposed remedies fail to consider the broader context. For example, while the United States unquestionably has the greatest scientific community in the world, the difficulties of administering basic research have increased due in good part to a failure to understand the difference between science and ordinary business activity. See Robert P. Crease and Nicholas P. Samios, "Managing the Unmanageable," *Atlantic* 267 (1991): 80–88. As they note, "at a time of widespread lamentations about the loss of U.S. technological competitiveness, it is ironic that we are destroying one of the most important means by which we established that technological competitiveness in the first place" (p. 87).

2. Of the many recent critiques of our universities, I think the best is still Allan Bloom, *The Closing of the American Mind* (New York: Simon and Schuster, 1986). Bloom is surely correct to stress the lack of focus of the contemporary curriculum, but even he does not explore with adequate depth the role of the natural sciences in our education and our civilization. For a lamentable example of "professor bash-

ing," in which a defense of traditional values is associated with ignorance of matters scientific, see Charles Sykes, *The Hollow Men* (Chicago: Regnery, 1990).

3. For a fuller discussion of the origins and formulation of the concept of a value-free science, see chap. 7 below.

4. The literature on this concept is extensive. For an introduction, see the thoughtful discussion in Michael Ruse, *Taking Darwin Seriously* (Oxford: Blackwell, 1986), chap. 6.

5. This volume was stimulated by the invitation to prepare a paper for the Claremont Institute Conference, "The Permanent Limits of Science," Feb. 15–16, 1991, at which I presented a shorter version of chaps. 2, 6, and 7. I am grateful to William Rusher and the participants at this conference for focusing on this topic.

6. *The Nature of Politics* (New Haven, Conn., Yale University Press, 1989).

7. What I have described as the "new naturalism" (ibid., esp. pp. 238–49) is, in many respects, similar to what has been called "critical naturalism" by Roy Bhaskar in *The Possibility of Naturalism: A Philosophical Critique of the Contemporary Human Sciences* (Atlantic Highlands, N.J.: Humanities Press, 1979) and *Scientific Realism and Human Emancipation* (London: Verso Books, 1986), as well as by Ian Shapiro in *Political Criticism* (Berkeley: University of California Press, 1990). See also Carl Degler, *In Search of Human Nature* (New York: Oxford University Press, 1991); Daniel C. Dennett, *Consciousness Explained* (Boston: Little, Brown, 1991); Lloyd Weinreb, *Natural Law and Justice* (Cambridge, Mass., Harvard University Press, 1987); Alisdaire MacIntyre, *After Virtue* (Notre Dame, Ind., University of Notre Dame Press, 1981).

8. As I will use the term "modern science," "modern" refers to the past 400 years of Western history, and "science" includes not only the natural sciences (physics, chemistry, biology, and the like) but also the extension of the methods of these disciplines to the study of human life (the social sciences, broadly so-called). As will be evident below (see chaps. 2 and 3), the human implications of scientific methods seem to depend less on the subject being studied than on the conception of science itself. Cf. Sir Francis Bacon, *The New Organon*, bk. 1, aphorism 127, in *Bacon: Selected Writings*, ed. Sidney Warshaft (New York: Odyssey, 1965), p. 379.

9. The primacy of utility and efficiency is obvious not only in our practical life but in such modern philosophic movements as utilitarianism: what is thought good is primarily what is useful. Although the concept of virtue is not totally ignored by modern philosophers, even these attempts have usually failed to do justice to the ancient view of this concept. See J. B. Schneewind, "The Misfortunes of Virtue," *Ethics* 101 (1990): 42–63; and Martha C. Nussbaum, *The Fragility of Goodness: Luck and Ethics in Greek Tragedy and Philosophy* (Cambridge: Cambridge University Press, 1986). On the relationship between religion, ancient philosophy, and modern philosophy as *the* fundamental question facing human thought in our civilization, see Laurence Berns, "The Relation between Philosophy and Religion," *Interpretation* 19 (1991): 43–60.

10. Throughout this book, I will use images or diagrams—like those a teacher puts on the blackboard—to illustrate relationships. On the role of images in human learning, see Plato, *Republic*, 4.420b, 6.488a, 7.514a, ed. Allan Bloom (New York: Basic Books, 1968), pp. 98, 167–68, 193.

11. This analysis reflects recent studies of the sense of justice from many scholarly disciplines: see Roger D. Masters and Margaret Gruter, eds., *The Sense of Justice: Biological Foundations of Law* (Newbury Park, Calif.: Sage, 1992). While

human social behavior cannot be reduced to that of chimpanzees or rhesus monkeys, the continuities between the nonhuman primates and our own species are obvious to anyone who has visited a zoo or seen television or film coverage of studies by Jane Goodall, Diane Fossey, Frans de Waal, or other primatologists. For examples, see Shirley C. Strum, *Almost Human: A Journey into the World of Baboons* (New York: W. W. Norton, 1987); Jane Goodall, *Through a Window* (Boston: Houghton Mifflin, 1990). Human politics and morality thus combine behaviors we share with primates with uniquely human attributes based on verbal language; see Glendon Schubert and Roger D. Masters, eds. *Primate Politics* (Carbondale: Southern Illinois University Press, 1991).

2. Modern Science

1. See, among others, Morris R. Cohen and Ernst Nagel, *An Introduction to Logic and Scientific Method* (New York: Harcourt Brace, 1934); Karl Popper, *The Logic of Scientific Discovery* (London: Hutchinson, 1959); Thomas Kuhn, *The Structure of Scientific Revolutions* (Chicago: University of Chicago Press, 1962); Paul Feyerabend, *Against Method: Outline of an Anarchistic Theory of Knowledge* (London: NLB; Atlantic Highlands, N.J.: Humanities, 1975), and *Farewell to Reason* (London and New York: Verso, 1987). For a survey of these issues as they relate to the social sciences, see May Broadbeck, ed., *Readings in the Philosophy of the Social Sciences* (New York: Macmillan, 1968), and Leonard I. Krimerman, ed., *The Nature and Scope of Social Science: A Critical Anthology* (New York: Appleton Century Crofts, 1969). Although much of the literature has been devoted to controversy between logical positivists and relativists of various sorts, the position usually described as realism (i.e., that humans can know how the world "really" is) is a distinct and valid understanding of the character of science; for an excellent statement of this view, see Richard W. Miller, *Fact and Method: Explanation, Confirmation and Reality in the Natural and Social Sciences* (Princeton, N.J.: Princeton University Press, 1987). On the relationship between contemporary evolutionary theory and science, see James H. Fetzer, ed., *Sociobiology and Epistemology* (Dordrecht: Reidel, 1985).

2. On reading the first draft of this essay, Michael Platt and John Scott independently recommended David Rapport Lachterman's *The Ethics of Geometry: A Genealogy of Modernity* (New York: Routledge, 1989). I found this book extraordinary and expect it will have a major effect in clarifying the philosophy and history of science. Lachterman's analysis of the difference between ancient and modern mathematics, science, and philosophy parallels my own presentation, despite the facts that he focuses on the philosophical foundations of mathematics (rather than a description of the scientific community) and explicitly denies an interest in defending or recommending "some kind of return" to ancient science (p. xiii). Given these differences in methods and objectives between Lachterman's account and my own, specialists who have questions about my understanding of the differences between the ancients and moderns may wish to consult his work. Another study of great importance in clarifying the difference between ancient and modern mathematics is Michael J. White's *The Continuous and the Discrete* (Oxford: Clarendon Press, 1992), which focuses on Aristotle's understanding of space and motion, rival views in antiquity, and the novelty of the modern conceptions underlying infinitesimal

calculus and Newtonian physics. These issues will be explored in detail in the next chapter.

3. Although "critical" philosophers of science like Feyerabend have emphasized the elements of bias and self-interest in the behavior of scientists, deception, self-deception, and venality can be found in any human activity. It is for this reason that one can probably gain a better understanding of the character of modern science by showing what it is *not*. This is particularly important because the question of whether humans could ever "know" anything without bias and self-interest is itself an issue within such scientific fields as cognitive neuroscience and physics. See below, chaps. 4 and 6.

4. Robert Pool, "Cold Fusion: Only the Grin Remains," *Science* 250 (1990): 754–55; John R. Huizenga, *Cold Fusion: The Scientific Fiasco of the Century* (Rochester, N.Y.: University of Rochester Press, 1992).

5. John H. Beckstrom, *Evolutionary Jurisprudence* (Champaigne–Urbana, University of Illinois Press, 1988), p. 39.

6. In classical Greek thought, a fundamental distinction was made between "the rules of nature" (*physis*) and "the rules of the laws," or conventions (*nomos*); in what is today called the dichotomy of nature versus nurture, the word *law* was entirely on the side of nurture, or human culture. See especially the famous fragment from Antiphon the Sophist, *On Truth*; cited in Roger D. Masters, "Nature, Human Nature, and Political Thought," in *Human Nature in Politics*, ed. Roland Pennock and John Chapman (New York: New York University Press, 1977), pp. 69–110. This usage characterizes the thought of Plato and Aristotle as well as of the pre-Socratics and Sophists generally. Hence, Plato and Aristotle typically use the phrase "according to nature" rather than "law of nature": e.g., Plato, *Republic*, 4.428e–429a; Aristotle, *Politics*, 1.2.1253a; *Nicomachean Ethics*, 5.7.1134b. I am indebted to Michael Rosano for indicating to me several important exceptions, see *Gorgias*, 83e; *Timaeus*, 483e. The one case of such a usage by Aristotle (*De Caelo*, 1.1.268a10–15) is clearly metaphorical: "And so having taken these three [sc., the three dimensions of space] from nature as (so to speak) laws of it, we make further use of the number three in the worship of the Gods." In the later Roman and especially the medieval Christian tradition, the concept of legal rights and duties recognized by all civilized peoples (*jus gentium*) led to the phrase *lex naturalis* (natural law) to describe moral obligations that all humans derive from being human. The most influential statement is doubtless St. Thomas Aquinas's "Treatise on Law" in the *Summa Theologica* 1–2, Q. 91–97. On the history of the concept of "law of nature," see John R. Milton, "The Origin and Development of the Concept of the 'Laws of Nature,'" *Archives européenes de sociologie* 22 (1981): 173–95; and Jane E. Ruby, "The Origins of Scientific 'Law,'" *Journal of the History of Ideas* 67 (1986): 341–60.

The combination of the terms *law* and *nature*, largely absent in antiquity before its use by Roman Stoics like Seneca, at first mainly referred to moral obligations. As Ruby shows, "law" (rather than "rule," "order," or "form") was systematically introduced to describe the regularities of optics by Roger Bacon (c. 1210–c. 1292); this sense of "law" was extended to physical nature by theologians like Ockham (c. 1285–c. 1349), whose thought was characterized by "a complete rejection of the metaphysical realism maintained by Aristotle and the earlier scholastics and an exceptionally strong emphasis on the absolute freedom and omnipotence of God, especially in relation to the works of creation" (Milton, p. 184). This theological tradition—described as a combination of nominalism ("the thesis that everything which exists is an individual") and voluntarism (with regard to nature, "the view that the world owes both its existence and its nature to a free choice of God")—differed from the thought of the ancients primarily on theological grounds, since "belief in a divine legislator for nature was not characteristic of Greek thought." (Milton, pp. 185–86). I am indebted to my colleague James Murphy for emphasizing the importance of this point and drawing my attention to the work of Ruby and Milton. For a more extensive treatment of the role of Christian theology at the foundations of the historical preconditions of modern science, see M. B. Foster, "The Christian Doctrine of Creation and the Rise of Modern Natural Science," *Mind* 43 (1934): 446–68; and Amos Funkenstein, *Theology and the Scientific Imagination from the Middle Ages to the Seventeenth Century* (Princeton, N.J.: Princeton University Press, 1986). The studies of Foster, Ruby, Milton, and Funkenstein are invaluable for any account of the historical origins of modern science and its difference from the science of antiquity.

In contrast to the ancients, Bacon transforms the Platonic concept of the Form of a thing into the modern concept of laws of nature, using the term *law* to describe the regularities and properties of the physical or biological world independently of human agency (Bacon, *New Organon*, 2.17, ed. Sidney Warshaft [New York: Odyssey, 1965], p. 388). Hobbes's use of the term *Law of Nature* (*Leviathan*, 1.14–15; ed. Schneider [New York: Library of Liberal Arts, 1958] pp. 109–32) is deliberately ambiguous; while the subject matter discussed concerns the *traditional* human actions covered by the term, Hobbes transforms it into a predictive and universally valid descriptive rule. Today, of course, the Baconian or modern usage has become so normal that we need to be reminded of its origins. For example, when Leonardo da Vinci describes natural necessity in the context of something like modern science, his use of term *rule* (*regola*) is translated as "law" by contemporary translators; e.g., "La neciessità è tema e inventrice della natura e freno e regola eterna" is rendered "Necessity is the theme and the inventress, the eternal curb and law of nature" (*The Notebooks of Leonardo da Vinci*, ed. Jean Paul Richter [New York: Dover, 1970], vol. 2, p. 285; "Necessity is the theme and inventor of nature, its eternal curb and law" *The Notebooks of Leonardo da Vinci*, ed. Irma A. Richter [Oxford: Oxford University Press, 1980], p. 7). Unless otherwise noted, for convenience references to Leonardo's *Notebooks* will be to the latter edition.

7. The classic statement is that of R. A. Fisher, "The Mathematics of a Lady Tasting Tea," in Fisher, *The Design of Experiments* (Adelaide: Hafner Press, Macmillan, 1971), pp. 11–26. On Fisher's reasons for developing statistical measurements in the study of human affairs, see chap. 6 below. For a careful assessment of the limits of this perspective, see Miller, *Fact and Method*.

8. José Ortega y Gasset, *The Revolt of the Masses* (New York: W. W. Norton, 1957), pp. 109–12. Cf. Bacon, *New Organon*, 1.122: "For my way of discovering sciences goes far to level men's wits, and leaves but little to individual excellence, because it performs everything by the surest rules and demonstrations" (ed. Warshaft, p. 369). For the ancient view, see Aristotle, *Metaphysics*, 2.1.993b.

9. Cf. Leo Strauss, *Persecution and the Art of Writing* (Glencoe, Ill.: Free Press, 1952).

10. Although many consider Kuhn's *The Structure of Scientific Revolutions* to be the standard account of this process, more recent work in the history and phi-

losophy of science has often qualified or challenged elements of this view; e.g., Donald T. Campbell, "Evolutionary Epistemology," in *The Philosophy of Karl Popper* ed. P. S. Schilpp (LaSalle, Ill.: Open Court, 1974), pp. 413–63. Cf. the references in n. 1, this chapter. While the rate of change in scientific theories or paradigms varies from slow accommodation to rapid acceptance, ultimate success is measured by the textbooks widely adopted for teaching science in public educational institutions. In any interpretation, modern science differs in this regard from the science of the ancients, organized as it was around individual teachers (Pythagoras, Socrates, Epicurus) or formal schools (the Academy, the Lyceum, the Stoa). No modern university is likely to confront the problems faced by the Lyceum when the original manuscripts of its founder, Aristotle, and his successor, Theophrastus, were removed as a protest over the election of the next head of the school. For the implications of this event, see Roger D. Masters, "The Case of Aristotle's Missing Dialogues," *Political Theory* 5 (1977): 31–60, and "On Chroust: A Reply," *Political Theory* 7 (1979): 545–57.

11. In many respects, the classic text remains Jacob Burkhardt, *The Civilization of the Renaissance in Italy* (1929; reprint, New York: Harper Torchbooks, 1958). For the historical antecedents of these developments, in addition to the works of Foster, Milton, Ruby, and Funkenstein cited in n. 6, this chapter, see Hans Blumenberg, *The Legitimacy of the Modern Age* (Cambridge, Mass.: MIT Press, 1983).

12. Leonardo is usually understood as an artistic genius and futuristic inventor (ignoring his role as a political advisor and innovative natural scientist and philosopher); Machiavelli is typically classified as a political thinker (ignoring both his active political career and his poetry and theater). More important perhaps, Machiavelli and Leonardo were friends, having met at the court of Cesare Borgia in 1502; as second secretary in the Florentine republic, Machiavelli was instrumental in securing Leonardo's commissions to paint the legendary and ill-fated *Battle of Anghieri* (his unfinished masterpiece, later destroyed) and to rechannel the Arno River to defeat Pisa. See *Notebooks of Leonardo*, pp. 348–56, and *The Unknown Leonardo*, ed. Ladislao Reti (New York: McGraw Hill, 1974), pp. 138–63. On the influence of Leonardo on Machiavelli's political thought, see Roger D. Masters, "Machiavelli and Hobbes: Theory and Practice at the Origins of Modernity" (Paper presented at the 1992 Annual Meeting of the American Political Science Association, Chicago, September 3–6, 1992).

13. *Notebooks of Leonardo*, p. 193. The *Notebooks* provide striking evidence of Leonardo's philosophic and scientific depth—and his radical departure from the procedures as well as the teachings of the medieval scholasticism, based on a return to Heraclitus and the Platonic concept of Form combined with a focus on the artist as creator or imitator with a power akin to that of the Judeo-Christian God. That his references to God should not be interpreted as evidence of traditional religious belief was evident in the first edition of Vasari's *Life of Leonardo*: "Leonardo was of so heretical a cast of mind that he conformed to no religion whatever accounting it perchance much better to be a philosopher than a Christian" (cited in *Notebooks*, p. 288).

14. Ibid., p. 6. On the relationship between technology and this scientific methodology, see n. 32, this chapter.

15. "Nessuna investigazione si può dimandare vera scienza, s'essa non passa per le mathematiche dimonstrazioni" (Leonardo da Vinci, *Trattato de la pintura*, cited in Ernst Cassirer, *Individual and the Cosmos in Renaissance Philosophy*, trans.

Mario Domandi [New York: Barnes and Noble, 1964], p. 154; *Notebooks*, p. 8). I am greatly indebted to Mr. John Scott, who first directed my attention to this citation and thereby to the importance of a deeper understanding of Leonardo's scientific method.

16. While this focus on pagan antiquity is the explicit theme of the *Discourses on Titus Livy*, it is equally evident in the examples and teaching of *The Prince*. See also *Mandragola*, Prologue: "In all things, the present age falls off from ancient *virtù*" (ed. Mera J. Flaumenhaft [Prospect Heights, Ill.: Waveland Press, 1988], p. 11). The plot of this comedy is perhaps the clearest manifestation of Machiavelli's intention to achieve what Nietzsche was later to call a "transvaluation of values."

17. *Discourses on the First Ten Books of Titus Livy*, bk. 1, Preface (ed. Bernard Crick [Harmondsworth, England: Penguin, 1970], p. 97). Cf. *Prince*, chap. 15: "I depart from the orders of others" (ed. Harvey Mansfield, Jr. [Chicago: University of Chicago Press, 1985], p. 61); *Mandragola*, Prologue: "We want you to understand a new case born in this city" [sc., Florence] (ed. Flaumenhaft, p. 9).

18. Cf. *Florentine Histories*, Preface (ed. Harvey Mansfield, Jr., and Laura Banfield [Princeton, N.J.: Princeton University Press, 1988], p. 7). In this sense, Machiavelli epitomized the Renaissance view that Burkhardt described as "the state as a work of art" (Burkhardt, *Civilization of the Renaissance*, pt. 1). Lest this seem a forced comparison, compare the following: "The painter is lord of all types of people and of all things. . . . If he wants valleys, if he wants from high mountain tops to unfold a great plain extending down to the sea's horizon, he is lord to do so; and likewise if from low plains he wishes to see high mountains" (Leonardo, *Notebooks*, pp. 194–95). "For just as those who sketch landscapes place themselves down in the plain to consider the nature of mountains and high places and to consider the nature of low places place themselves high atop mountains, similarly, to know well the nature of peoples one needs to be prince, and to know well the nature of princes one needs to be of the people" (Dedication, *The Prince*, ed. Harvey C. Mansfield, Jr. [Chicago: University of Chicago Press, 1985]), p. 4.

19. *Prince*, chap. 24: "It is not unknown to me that many have held and hold the opinion that worldly things are so governed by fortune and by God, that men cannot correct them with their prudence. . . . Nonetheless, in order that our free will not be eliminated, I judge that it might be true that fortune is arbiter of half of our actions, but also that she leaves the other half, or close to it, for us to govern" (ed. Mansfield, p. 98). Hence, for Machiavelli, the best regime need *not* be located where the natural resources are favorable; instead, he suggests the advantages of an unfavorable natural site in which the hostile effects of the environment are controlled by law. Nature is a raw material to be shaped by human endeavor, not a standard of judgment (as for Plato or Aristotle) or a symbol of corruption (as for Paul and the early Church Fathers). As further evidence, compare Machiavelli's image of fortune as "one of those violent rivers" that cause untold damage "when they become enraged" (ibid.) with Leonardo's assertion that "amid all the causes of destruction of human property, it seems to me that rivers hold the foremost place on account of their excessive and violent inundations" (Leonardo, *Notebooks*, p. 23). It does not seem an accident that Leonardo attempted to rechannel the Arno so that Florence could defeat Pisa, while Machiavelli illustrated his political theory by the allegory of "dams and dikes" that could channel the "river" of "fortune."

20. "Nature to be commanded must be obeyed" (Bacon, *New Organon*, 1.3 [ed. Warshaft, p. 331]).

21. Ibid., 1.124 (pp. 372–74); 2.1 (p. 376). This is possible because what we call physics (in Bacon's terms, "natural philosophy") can transform *all* science: "astronomy, optics, music, a number of the mechanical arts, medicine itself—nay, what one might more wonder at, moral and political philosophy and the logical sciences" (1.80 [p. 353]).

22. "Only let the human race receive that right over nature which belongs to it by divine bequest, and let power be given to it; the exercise thereof will be governed by sound reason and true religion" (ibid., 1.124 [p. 374]). See also Machiavelli's critique of both ancient philosophy and Christianity on the ground that both focus on "what should be done" rather than the "effectual truth" of "what is done" (*Prince*, chap. 15; ed. Mansfield, p. 61). It follows that modern scientific and technological methods are—at least in principle—available to all rulers, regardless of the virtue of their character or the morality of their objectives: although Leonardo kept the invention of the submarine in cipher, he worked in the service of Cesare Borgia until the "accident" that cost Cesare his power, and after working for the Florentine republic he was in the service of Milan and then of France. Cf. Burkhardt, *Civilization of the Renaissance*, vol. 1, pp. 59, 127–32, with Machiavelli, *Prince*, chap. 7 (ed. Mansfield, p. 32 and n. 14).

23. In understanding this change, Lachterman's *Ethics of Geometry* is invaluable. He shows, for example, how Euclidian (ancient) geometry resisted the modern notion that ratios and numbers are homogeneous (pp. 33–41). Descartes, Hobbes, and other early modern geometers laid the foundations of the calculus by asserting this homogeneity. For the ancients, a ratio (or *logos*) is not a number; the heterogeneity of "natural" kinds lies at the root of all mathematical thinking. For the moderns, in contrast, all mathematical notations are ultimately operators; hence, for the moderns, it is a convention whether the series is noted in fractions or in decimals. On the implications of this difference, see chap. 3.

24. See René Descartes, *Discourse on Method*, esp. pt. 6: "[I]t is possible to attain knowledge which is very useful in life, and . . . instead of that speculative philosophy which is taught in the Schools, we may find a practical philosophy by means of which, knowing the force and action of fire, water, air, the stars, heavens and all other bodies that environ us, as distinctly as we know the different crafts of our artisans, we can in the same way employ them in all those uses to which they are adapted, and thus render ourselves the masters and possessors, as it were, of nature" (trans. Paul J. Olscamp [Indianapolis: Bobbs-Merrill Co., 1965], p. 50). Cf. Bacon, *New Organon*, 1.122 (ed. Warshaft, p. 369), cited below. Lachterman provides a powerful account of how the *creative* or *constructive* character of modern mathematics lies at the foundation of these utilitarian or technological consequences of science. On the medieval precursors of this transformation, see Funkenstein, *Theology and the Scientific Imagination*.

25. *Leviathan*, Author's Introduction, chap. 13. In ancient Greece the introduction of geometry to political thought was associated with the skepticism of Sophists like Gorgias (who used the difference between Euclidean geometry and political rhetoric to demonstrate that all human practices were conventional). Hobbes accepts the notion that laws are customs or conventions, based on a social contract, but transforms it by assuming that this discovery, based on natural science, can be used as a guide to constructing a better political order. Cf. Masters, "Machiavelli and Hobbes."

26. Even Hobbes's famous picture of the "natural condition" of mankind as a

"war of all against all" implies—if the negatives are deleted—that proper political institutions can produce a world in which "there is . . . place for industry, because the fruit thereof is . . . certain; . . . culture of the earth, . . . navigation, . . . use of the commodities that may be imported by sea; . . . commodious building, . . . instruments of moving and removing such things as require much force, . . . knowledge of the face of the earth, . . . account of time, . . . arts, . . . letters, . . . society" (*Leviathan*, pt. 1, chap. 13 [ed. Schneider, Library of Liberal Arts, p. 107]). On Hobbes as the great teacher of civilization and civil society based on science and technology, see Heinrich Meier, *Carl Schmitt, Leo Strauss, et la notion de la politique* (Paris: Fayard, 1990), pp. 55–61, 88, 118–19.

27. Although Hobbes's *Leviathan* seemingly emphasizes an absolute sovereign, the work had already established the notion of representative government. For Hobbes, the "authorities" in a commonwealth are the "representative" of the multitude, whose consent alone can "authorize" or give a "commission to act" (*Leviathan*, pt. 2, chap. 16 [p. 135]). Locke, of course, makes the notion of the "consent" of the governed into the central concept of liberal or constitutional theory: *Two Treatises of Government*, esp. *Second Treatise* (chap. 8, sec. 95–98; ed. Thomas I. Cook [New York: Hafner, 1956], pp. 168–70).

28. The writings of French *philosophes*—notably the *Encyclopédie* of Diderot and d'Alembert—symbolize the combination of mathematics, natural science, technological innovation, and political reformism that epitomizes the modern scientific project. See n. 31, this chapter.

29. In this regard, it is worth emphasizing that Newton and Locke were not only personal friends but both officials in the constitutional regime established in England by the Glorious Revolution of 1688. My historical précis of the origins of modern science has quite intentionally embedded questions of natural science in a broader context of politics and philosophy. In the deepest sense, the unfettered search for the "laws of nature" was impossible until societies were governed by constitutional law rather than by divine right or arbitrary will. Since the freedom and rule of law that seem necessary for the openness and skepticism of modern science and technology are very rare, perhaps it should be no surprise that this approach to knowledge did not emerge elsewhere in human history.

30. For example, Arnold Brecht has summarized the scientific method as having eleven "scientific actions" or "scientific operations": "1. *Observation*, . . . 2. *Description* . . . 3. *Measurement* . . . , 4. *Acceptance* or nonacceptance (tentative) as *facts* or *reality* of the results of observation, description, and measurement. 5. *Inductive generalization* (tentative) . . . 6. *Explanation* (tentative) . . . 7. *Logical deductive reasoning* from inductively reached factual generalizations (no. 5) or hypothetical explanations (no. 6) . . . 8. *Testing* . . . 9. *Correcting* . . . 10. *Predicting* events or conditions to be expected as a consequence of past, present, or future events or conditions . . . 11. *Nonacceptance* (elimination from acceptable propositions) of all statements not obtained or confirmed in the manner here described" (*Political Theory* [Princeton, N.J.: Princeton University Press, 1968], pp. 28–29).

31. In addition to his experience as a military advisor to the Florentine republic, Cesare Borgia, the Duke of Milan, Giuliano de' Medici, and the French King François I, (e.g., *Notebooks*, pp. 338, 344, 350–52), Leonardo thought of numerous useful "instruments of war" (pp. 294–97). His idea for a submarine (pp. 96–97) is well known, as is his attempt to construct an airplane (pp. 103–6, 298). Perhaps more illustrative of the breadth of modern technology are Leonardo's ideas

for the modern house (pp. 214–15), towns with overpass highways (pp. 213–14), running water both inside the home and for the community as a whole (pp. 215–16; 326, 370, 386–87), a method for minting standardized coins of high quality (pp. 379–80), and a science of physics capable of producing these advances in applied mechanics (pp. 55–86). Cf. Reti, *Unknown Leonardo*, pp. 190–215, 240–91.

32. From the outset, it was evident that technological innovations could contribute to the pure sciences; a century before Galileo, Leonardo tried to "construct glasses in order to see the moon large" (*Notebooks*, pp. 52, 378–79). Although he does not seem to have been successful in actually making a telescope, Leonardo did construct a device with which he could watch a solar eclipse without damage to his sight (p. 298). In contrast, Aristotle asserts explicitly that the innovations "required" for civilization were already known by his time (*Politics*, 7.10.1130b); the only technological invention attributed to Aristotle is an alarm clock to permit him to extend the time he devoted to scientific research and philosophical contemplation.

33. For the argument that this characteristic is true of the most basic of the innovations on which civilization rests, namely writing, see "A Writing Lesson" in Claude Lévi-Strauss, *Tristes Tropiques* (New York: Atheneum, 1974). For present purposes, it is indifferent whether all technological innovations are value-neutral in a sense—and modern technology merely exaggerates this tendency—or rather that our own inventions are peculiarly open to a dissociation between the machine and the uses to which it is put.

34. The classic definition of this meaning of democracy is presented in bk. 8 of Plato's *Republic*: a democratic regime is "like a many-colored cloak decorated in all hues, . . . decorated with all dispositions" (557c), while a democratic individual "lives his life in accord with a certain equality of pleasures he has established" (561b).

35. The premise that social organization should promote personal freedom is explicit for most modern thinkers, though for some (e.g., Locke), the individual's free choice lies at the origin of political institutions; whereas for others (e.g., Marx) it is only possible at the end of history. This emphasis on individual freedom or independence seems to have been derived from the quest for political freedom that was evident at the outset of modernity, not only in Machiavelli (e.g., *Discourses on Titus Livy*, 1.10) but also in Leonardo: "When besieged by ambitious tyrants I find a means of offence and defence in order to preserve the chief gift of nature, which is liberty" (*Notebooks*, p. 284). Even Rousseau, who of the moderns was most insistent on the need to restore the ancient primacy of virtue, based his political teaching on the natural equality and freedom of all men; cf. *Discourse on Sciences and Arts*, or *First Discourse* (in *Collected Writings of Rousseau*, ed. Roger D. Masters and Christopher Kelly [Hanover, N.H.: University Press of New England, 1992], vol. 2, pp. 1–22) with *Discourse on the Origins of Inequality*, or *Second Discourse*, pt. 1 (in *Collected Writings of Rousseau*, vol. 3, pp. 1–95]).

36. Hobbes, *Leviathan*, Author's Introduction:
. . . whosoever looks into himself and considers what he does when he does *think, opine, reason, hope, fear*, etc., and upon what grounds, he shall thereby read and know what are the thoughts and passions of all other men upon the like occasions. I say the similitude of *passions*, which are the same in all men: *desire, fear, hope*, etc.; not the similitude of the *objects* of the passions, which are the things *desired, feared, hoped*, etc., for these the constitution individual and particular education

do so vary, and they are so easy to be kept from our knowledge, that the characters of man's heart, blotted and confounded as they are with dissembling, lying counterfeiting, and erroneous doctrines, are legible only to him that searches hearts. (ed. Schneider, p. 25)

37. See Ithiel de Sola Pool, ed., *The Social Meaning of the Telephone* (Cambridge, Mass.: MIT Press, 1977). Interestingly enough, the critical invention seems to have been the telephone switchboard, not the telephone itself. The obvious component of a technology is not always its most important one.

38. The literature on this problem is extensive. In many respects the classic remains C. P. Snow's *Science and Government* (Cambridge, Mass.: Harvard University Press, 1956). For a suggested method of addressing this problem, see Roger D. Masters and Arthur Kantrowitz, "Scientific Adversary Procedures: The SDI Experiments at Dartmouth," in *Technology and Politics*, ed. Michael Kraft and Norman Vig (Durham, N.C.: Duke University Press), pp. 278–305.

39. For an introduction, see David Barash, *Sociobiology and Behavior*, 2nd ed. (New York: Elsevier, 1982). The varied patterns among birds are well documented in Paul Ehrlich, David S. Dobkin, and Darryl Wheye, *The Birder's Handbook: A Field Guide to the Natural History of North American Birds* (New York: Simon & Schuster, 1988). Using this book and a bird identification guide, anyone who enjoys observing wildlife can learn to see at firsthand the importance of social behavior among other species.

40. In addition to the citations in the preceding note, see Robert L. Trivers, *Social Evolution* (Menlo Park, Calif.: Benjamin/Cummings, 1985); and, for human applications, Eric Alden Smith and Bruce Winterhalder, eds., *Evolutionary Ecology and Human Behavior* (New York: Aldine de Gruyter, 1992). As critics have pointed out, it is often difficult to ascertain the precise environmental factors or selective advantages that explain observed variations; see Philip Kitcher, *Vaulting Ambition* (Cambridge, Mass.: MIT Press, 1985). Such considerations are, however, characteristic of the debates in the actual process of modern scientific research rather than evidence against the existence of the phenomena themselves. And since much behavioral variability is due to individual learning or social adaptation, the habitual criticisms of "sociobiology" as a form of genetic determinism are irrelevant to an assessment of behavioral ecology. See especially Richard Alexander, *Darwinism and Human Affairs* (Seattle: University of Washington Press, 1979).

41. For an exceptionally thorough survey of the variations in human mating and marriage, showing not only the explanatory power of cost–benefit analysis but the value of its use in policy-making and jurisprudence, see Richard Posner, *Sex and Reason* (Cambridge, Mass.: Harvard University Press, 1992). On the theoretical relationship between economic analysis and evolutionary biology, see Jack Hirshleifer, "Economics from a Biological Viewpoint," *Journal of Law and Economics* 20 (1977): 1–52, and *Economic Behavior in Adversity* (Chicago: University of Chicago Press, 1987).

42. E.g., Marvin Harris, *Of Cannibals and Kings* (New York: Random House, 1977); Pierre van den Berghe, *Human Family Systems* (New York: Elsevier, 1979); Napoleon Chagnon and William Irons, eds., *Evolutionary Biology and Human Social Behavior: An Anthropological Perspective* (North Scituate, Mass.: Duxbury Press, 1979); Roger D. Masters, "Explaining 'Male Chauvinism' and 'Feminism': Cultural Differences in Male and Female Reproductive Strategies," in *Biopolitics and Gender*, ed. Meredith Watts (New York: Haworth Press, 1984), pp. 165–210,

and *The Nature of Politics* (New Haven, Conn.: Yale University Press, 1989), esp. chaps. 1 and 6.

3. Ancient Science

1. [T]he description of the number line and its arithmetic is at the root of our understanding and intuition about mathematics today. . . . Greek mathematics up to the second century BC seems, to an extraordinary degree, to be different. . . . Geometry, modelled on the properties of the three-dimensional space we seem to experience about us, provides the main ingredient of Greek mathematics" (D. H. Fowler, *The Mathematics of Plato's Academy* [Oxford: Clarendon Press, 1987], p. 10). This work is invaluable as a reconstruction of Greek mathematics, showing precisely how their way of thinking differed from modern "arithmetical" (or algebraic) mathematics. As late as the middle of the eighteenth century, the specialist we today call a mathematician was still called a geometer; e.g., Rousseau, *First Discourses* (in *Collected Writings of Rousseau*, ed. Roger D. Masters and Christopher Kelly [Hanover, N.H.: University Press of New England, 1992], Vol. 2, pp. 12, 19, 21). Interestingly enough, Leonardo's notebooks show not only a careful study of Euclidian geometry and the mathematics of the ancients but an interest in texts on mathematics that explored the discovery of "algebra" by Arabic mathematicians: e.g., *The Notebooks of Leonardo da Vinci*, ed. J. P. Richter (New York: Dover, 1970), vol. 2, p. 449.

2. In addition to the importance of inertia in the history of physics, the concept was seized upon by Hobbes as the physical analogue of the human desire for self-preservation (Hobbes, *Leviathan*, pt. 1, chap. 1–3, ed. Herbert W. Schneider [New York: Library of Liberal Arts, 1958] pp. 25–36).

3. See Jacques Ninio, *L'empreinte des sens* (Paris: Odile Jacob, 1989), chap. 1.

4. If the trajectory of cannonballs, reflecting military technology, provided one impetus to this development, the spread of commerce was another. More and more work in daily life required counting, adding, subtracting, multiplying, dividing. A mechanical device capable of performing these operations could contribute as much to peaceful arts as the discovery of ballistic trajectories could add to military ones. The importance of the shift from geometric forms to algebraic equations is underlined by the fact that two of the greatest mathematical geniuses of the seventeenth century—Pascal (1623–1662) and Leibniz (1646–1716)—were among the first to construct workable mechanical computing devices. For the first time, mathematicians began to think in terms of the algorithm, or operational rule, that could produce any outcome. Such a task is self-evident in constructing a mechanical calculating machine, since wheels and gears move continuously; the numbers in the dials can produce a continuous series of numbers as the outcome. See J. A. V. Turk, *Origin of Modern Calculating Machines* (Chicago: Western Society of Engineers, 1921). By the time Newton focused on the completion of the Galilean revolution in physics, scientists were recognizing the utility of understanding nature and its mathematical representation in terms of continuous motion or change.

5. See Freidrich Solmsen, "Nature as Craftsman in Greek Thought," *Journal of the History of Ideas* 24 (1963):473–96. Although the pre-Socratics' nonteleological view of *physis* might be viewed as an exception, even the pre-Socratics linked nature as "process" with its results or visible consequences in a way distinct from the char-

acteristic presuppositions of modern natural science. Cf. William Arthur Heidel, "*Peri physeos:* A Study of the Conception of Nature among the Pre-Socratics," *Proceedings of the American Academy:* 45 (1910):79–133. I am indebted to James B. Murphy for directing my attention to these sources.

6. See David R. Lachterman, *Ethics of Geometry: A Genealogy of Modernity* (New York, Routledge, 1989), esp. pp. 33–41; Fowler, *Mathematics of Plato's Academy,* chap. 7 et passim. Those who had difficulty with fractions in school will be surprised to learn that their reaction reflects the hesitancy of the ancients before the typically modern solution to one of the deepest problems of human reason. The resulting difference between ancients and moderns is symbolized by what is called Zeno's paradox. For the ancients, it is impossible to explain why the fleet-footed Achilles could pass the slow Tortoise once the latter had a head start, since during each moment of time, each moved a portion of the remaining distance toward the goal. At first sight puzzling, the principle behind this paradox is fundamental to the emergence of modernity. According to the mathematical principles dominant in antiquity and articulated by Aristotle, continuous space (measure) is not reducible to number. Zeno's paradox thus concerns ratios, and a distance can be divided an infinite number of times without fully measuring its extent: that is, one can divide a line in half, then again in half, and repeat the process indefinitely without reaching the end. See Michael J. White, *The Continuous and the Discrete* (Oxford: Clarendon Press, 1992), esp. chaps. 1–3 and cf. *De Caelo,* 1.1.268a–268b10. Although some ancients (Epicureans and Stoics) dissented from this understanding, not until modernity did mathematicians develop the notion of points without spatial measure that could be added to create a continuous line or surface; in this modern view, a measurable mathematics of continuous motion or change, represented by infinitessimal calculus, could become the basis of a new physics. For a popular introduction to this paradox, see Douglas Hofstader, *Gödel, Escher, Bach: An Eternal Golden Braid* (New York: Basic Books, 1979). ADDED IN PRESS: Professor John Scott has just kindly indicated to me Jacob Klein's "The World of Physics and the 'Natural' World," in *Lectures and Essays,* ed. Robert B. Williamson and Elliott Zuckerman (Annapolis, MD: St. John's College Press, 1985), pp. 1–35. Originally given at the University of Marburg in February, 1932, this essay (translated by David Lachterman) is of exceptional importance in explaining the origins and depth of the difference between ancient and modern science.

7. Carl Hempel, "The Function of General Laws in History," *Journal of Philosophy* 39 (1942):35–48 (reprint, pp. 231–32). For a careful critique, see Richard W. Miller, *Fact and Method: Explanation, Confirmation and Reality in the Natural and Social Sciences* (Princeton, N.J.: Princeton University Press, 1987), esp. chap. 1.

8. Not being a mathematician, yet seeing (as did thinkers as diverse as Aristotle, Hobbes, and Rousseau) the essential role of mathematical forms in philosophic discourse, I have relied heavily on the popularization in James Gleick, *Chaos* (New York: Viking, 1987). See also H.-O. Pietgen and R. H. Richter, *The Beauty of Fractals* (New York: Springer-Verlag, 1986), and for surveys of scientific work revealing the central role of chaos, David Ruelle, *Chance and Chaos* (Princeton, N.J.: Princeton University Press, 1991), as well as Saul Krasner, ed., *The Ubiquity of Chaos* (Washington, D.C.: American Association for the Advancement of Science, 1990). On the role of such processes in the emergence of new forms, see Per Bak and Kan Chen, "Self-Organized Criticality," *Scientific American* 264 (1991):46–53.

9. John Horgan, "Quantum Philosophy," *Scientific American* 26 (1992):5, 94–

104. While associated with twentieth-century physics, some of these basic concepts had already been crudely suggested in antiquity. The possibility of an effect of observation on the phenomena being observed was implied by some of the pre-Socratics (e.g., Protagoras), while the Epicureans challenged the dominant Aristotelian view of continuity by proposing something akin to discontinuous quanta (White, *The Continuous and the Discrete*, esp. chaps. 5–6).

10. The standard source is Hempel, "The Function of General Laws in History." For further references, see n. 1, chap. 2, above.

11. Perhaps the most impressive example concerns the origins and history of the kind of astronomic motion traditionally explained by Newtonian physics; see William Tittemore, "Chaotic Motion of Europa and Ganymede and the Ganymede-Callisto Dichotomy," *Science* 250 (1990):263–67; and Gerald Jay Sussman and Jack Wisdom, "Chaotic Evolution of the Solar System," *Science* 256 (1992): 56–62. Other examples, as Gleick points out, range from the shapes of snowflakes to the patterns of turbulent flow, weather, and plant growth. In human affairs, some researchers now claim to find chaotic processes in the course of epidemics, the pattern of economic change, and even the workings of the human brain.

12. On self-similarity, see Gleick, *Chaos*, pp. 96–118. When Aristotle claims that a line can be divided an infinite number of times without exhausting its measure, he means that the division of the last remaining segment in halves creates the same ratio as the first such division (see White, *The Continuous and the Discrete*, chap. 4 et passim). For the equivalent in fractal geometry, see Gleick, *Chaos*, pp. 98–102.

13. See the striking images in Gleick, *Chaos*, or Pietgen and Richter, *The Beauty of Fractals*, as well as the computer programs described in *Scientific American* and other popular science magazines.

14. Walter J. Freeman, "The Physiology of Perception," *Scientific American* 264 (1991):78–85.

15. See James Barham, "A Poincaréan Approach to Evolutionary Epistemology," *Journal of Social and Biological Structures* 13 (1991):193–258.

16. Ian A. McLauren, ed., *Natural Regulation of Animal Populations* (New York: Atherton, 1971); Edward O. Wilson, *Sociobiology* (Cambridge, Mass.: Harvard University Press, 1975), p. 81.

17. Stephen Jay Gould, *Wonderful Life* (New York: Norton, 1989).

18. Albert Somit and Steven A. Peterson, eds., *The Punctuated Equilibrium Debate: Scientific Issues and Implications*, Special Issue of *Journal of Social and Biological Structures*, vol. 12, no. 21–23 (1989), republished in expanded form as *The Dynamics of Evolution* (Ithaca, N.Y.: Cornell University Press, 1992).

19. Plato's formulation of the mathematics of population size is as follows:

. . . bearing and barrenness of soul and bodies come not only to plants in the earth but to animals on the earth when revolutions complete for each the bearing round of circles; for ones with short lives, the journey is short; for those whose lives are the opposite, the journey is the opposite. . . . For a divine birth there is a period comprehended by a perfect number; for a human birth, by the first number in which root and square increases, comprising three distances and four limits, of elements that make like and unlike, and that wax and wane, render everything conversable and rational. Of these elements, the root four-three mated with the five, thrice increased, produces two harmonies. One of them is equal an equal number of times, taken one hundred times over. The other is of equal length in one way but is an oblong; on one side, of one hundred rational diameters of the five, lacking one for each; or, if of irrational diameters, lacking two for each; on the other side, of one hundred cubes of the three. This whole geometrical number is sovereign of better and worse begettings. (*Republic*, 7.546b–d, ed. Allan Bloom [New York: Basic Books, 1968], p. 224). Cf. Bloom's note, pp. 467–68.

20. Compare Stephen Hawking, *A Brief History of Time* (New York: Bantam, 1988) with Roger Penrose, *The Emperor's New Mind* (New York: Oxford University Press, 1989), and Werner Heisenberg, *Physics and Philosophy* (New York: Harper Torchbooks, 1958).

21. On the substance of Plato's physics, see Paul Friedländer's remarkable "Plato as Physicist," in Friedländer, *Plato: An Introduction*, vol. 1; Bollingen Series, no. 59 (New York: Pantheon, 1958), pp. 246–60. I am indebted to James B. Murphy for bringing this valuable text to my attention.

22. On the level of organism and development, see Gerald Edelman, *Topobiology* (New York: Basic Books, 1988); for the problem as it applies to protein molecules, see Frederic M. Richards, "The Protein Folding Problem," *Scientific American* 264 (1991): 54–65.

23. As Friedländer puts it, Plato transcends the debate over the priority of causality or chaos that became central in Western thought after Hume: "Nature is both strict mathematical law and chaotic chance; chance dwelling in the realm of the absolutely indeterminate; law supervening in the form of mathematical order, upon the chaotic disorder without ever being able to control the latter completely. Thus neither law nor chance reigns supreme, but the world as we know it is a product of both" (*Plato: An Introduction*, vol. 1, p. 258). It need only be added that the word *law* is probably inappropriate as a description of Plato's view of order, or Form: see n. 6, chap. 2, above.

24. In pt. 3 I show that the conventional argument against the "naturalistic fallacy" ultimately rests either on the deduction of values from facts or the assertion of arbitrary standards without basis in either logic or fact. First, however, it is necessary to explain how the ancients themselves related naturalism and teleological reasoning (the remainder of this chapter) and to consider challenges to the very possibility of knowledge about human life (pt. 2).

25. Heraclitus, frag. 228, in *The Presocratic Philosophers*, ed. G. S. Kirk and J. E. Raven (Cambridge: Cambridge University Press, 1957), p. 202; Aristotle, *Meteorologica*, 2.2.355a13. Cf. Plato, *Republic*, 6.498a (ed. Bloom, p. 177). For evidence that the "one and the many" is the central issue of ancient social thought, see Arlene W. Saxonhouse, *Fear of Diversity: The Birth of Political Science in Ancient Greek Thought* (Chicago: University of Chicago Press, 1992). Note that, in modern mathematics and science, this problem is often assumed out of existence. If 3.14, 3.14159, and 3.1416 are equivalent representations of π, the "many" values (depending on the precision of measurement) can be treated as one and the same without reference to differences among them. See p. 33.

26. Paul Friedländer, "Plato as Geographer," in *Plato: An Introduction*, vol. 1, pp. 261–85. On the role of testing sense impressions in ancient science, see Ninio, *L'empreinte des sens*, chap. 1.

27. Heraclitus, frag. 222, in *The Presocratic Philosophers*, ed. G. S. Kirk and J. E. Raven (Cambridge: Cambridge University Press, 1957), p. 199. "The views of modern physics are in this respect very close to those of Heraclitus if one inter-

prets his element fire as meaning energy" (Heisenberg, *Physics and Philosophy*, pp. 70–71).

28. This should hardly be surprising; according to Aristotle (who was in a position to know), Plato "having in his youth first become familiar with Cratylus and with the Heraclitean doctrines (that all sensible things are ever in a state of flux and there is no knowledge about them), these views he held even in later years" (Aristotle, *Metaphysics*, 1.6.987a–b).

29. Plato, *Phaedrus*, 270d (*Collected Writings of Plato*, ed. Edith Hamilton and Huntington Cairns [Princeton, N.J.: Princeton University Press, 1961], p. 516).

30. Plato, *Republic*, 4.436c.

31. See the references in n. 20, this chapter. Much of the wonder of contemporary cosmology arises from the radical disjunction between our everyday life and the highly mathematical, counterintuitive pictures of the origin of the universe—a disjunction that is easier for most people to translate into such fancies of science fiction as "Star Trek" than to understand in technical terms.

32. For example, the jerky or "stick-slip" motion of two solid surfaces, which creates a squeaky hinge, cannot be explained solely on the basis of classical dynamics since the motion depends on the geometry of the molecules at the interface of the two surfaces and not merely on friction as an abstract concept. See Peter A. Thompson and Mark O. Robbins, "Origin of Stick-Slip Motion in Boundary Lubrication," *Science* 250 (1990): 792–94.

33. Plato, *Republic*, 3.414d–415d (the "noble lie"), 5.450d–451b, 452b–d, 454c, 473c; 9.592a–b; Aristotle, *Politics*, 2.2–5.1261b–1264b. See also Allan Bloom, "Interpretive Essay," *Republic*, pp. 380–91, and, more generally, Saxonhouse, *Fear of Diversity*.

34. *Metaphysics*, 4.5.1008b; *Collected Works of Aristotle*, vol. 2, p. 1592. It is probably for this reason that Aristotle—who ought to be our most reliable commentator in this regard—describes Plato's philosophy as in similar terms: *Metaphysics*, 4.2.1004b.

35. Although the ancient emphasis on the form or essence of a species seems to contradict this view, this follows only if the modern definition of nature is read back into their texts. For Aristotle, "since the ultimate species are substances and individuals which do not differ in species are found in them (e.g., Socrates, Coriscus), we must either describe the universal attributes first or else say the same thing many times over"; this focus on "the universal attributes of the groups that have a common nature" can thus be applied in comparing different kinds of animals as well as in identifying a single species. (*Parts of Animals*, 1.4.644a; *Collected Works of Aristotle*, vol. 1, p. 1002. As a result, even "higher" or taxonomic groups of species that "have certain attributes in common" are natural units if they "only differ in degree, and in the more or less of an identical element that they possess" (ibid., p. 1001). Aristotle contrasts such related groups, whose names are viewed as natural, with those "whose attributes are not identical but analogous"; contemporary biologists make the same distinction between homologous and analogous comparisons, pointing to the former as reflecting common ancestry and hence "natural" grouping. See Mario von Cranach, ed., *Methods of Comparing Animal and Human Behavior* (The Hague: Mouton, 1976).

36. See Martha C. Nussbaum, *The Fragility of Goodness* (Cambridge: Cambridge University Press, 1986).

37. A contemporary reassessment of this concern, which can be found in the

works of thinkers from Plato and Aristotle to Rousseau and Hegel, is the focus of my forthcoming *The Nature of Obligation*.

38. William D. Hamilton is now developing mathematical models of the relationships between parasites and hosts that show how sexual reproduction may have evolved as a means of avoiding linear or deterministic relationships, which can often lead the host species (and hence, ultimately, the parasite) to become extinct. Sexual recombination having evolved, changes in the populations of hosts and parasites are often modeled by a strange attractor, with no determinate solution. "The Uses of Biology in the Study of Law," (Lecture presented to Gruter Institute Seminar, Olympic Valley, Calif., June 21, 1992).

39. Roger D. Masters, "Classical Political Philosophy and Contemporary Biology," in *Politikos*, ed. Kent Moors (Pittsburgh: Duquesne University Press, 1989), vol. 1, pp. 1–44.

40. Thucydides, *History of the Pelopponesian War*, esp. the "archeology"; Plato *Republic*, 7.545d–547c, 10.614b–621d ("the Myth of Er"); Aristotle *Politics*, 4.1.1301a20 ff. Cf. Mircea Eliade, *Cosmos and History* (New York: Harper, 1959).

41. Locke, *Essay on Human Understanding*, 1.2.5 (see "Locke and the Origins of Value-Free Science," in chap. 7, below). See also Hobbes, *Leviathan*, esp. 1.11.

4. Is Knowledge Possible?

1. Barry Barnes and David Bloor, cited in Larry Laudan, *Science and Relativism: Some Key Controversies in the Philosophy of Science* (Chicago: University of Chicago Press, 1990), p. 106.

2. See Paul Feyerabend, *Against Method: Outline of an Anarchistic Theory of Knowledge* (London: NLB; Atlantic Highlands, N.J.: Humanities, 1975), and *Farewell to Reason* (London and New York: Verso, 1987); David Berlinski, *Black Mischief: Language, Life, Logic, Luck*, 2nd ed. (Boston: Harcourt Brace Jovanovich, 1988), and for a fuller discussion of epistemological questions, chap. 7, below. The position that science or objective truth rests on knowable principles is sometimes described as foundationalism (if focused on the principles), sometimes as realism (when the issue is the congruence of human knowledge and "reality"), and sometimes simply as scientific objectivity. To pose the problem clearly, however, it is dangerous to begin from specialized definitions that are difficult for nonspecialists to understand and not universally accepted even among contemporary philosophers.

3. That postmodernism and extreme skepticism have flourished in the humanities is perhaps not surprising, for these fields, which once led our universities in prestige and popularity, have become highly controversial over the past decade. Many of the criticisms of the contemporary university have been predominantly if not entirely directed at literature, philosophy, and other humanistic disciplines: see Dinesh D'Souza, *Illiberal Education* (New York: Free Press, 1991) as well as the works cited in chap. 1, n. 2 above. Critics of postmodernism complain, not unreasonably, that humanists have substituted "politically correct" interpretation for the proper respect for the traditional "canon" of Great Books. Too little attention has been given, however, to the reasons for these trends. The simple fact is that, compared to other fields, there are declining enrollments and fewer good students in the humanities. Television has ravaged the capacity of our students to read difficult

books, not to mention its effects on writing skills, and preprofessionalism has led to flooding enrollments in the social sciences. Under these circumstances, it should not be surprising that the many humanists with traditional approaches to teaching and scholarship are not as influential as they once were or that ideological appeals have been used by some in an attempt to attract students and attention.

4. Consider the following survey of theoretical answers to the problem of irreconcilable moral conflicts written by a critic of relativism:
Most nonrelativistic theories in the history of ethics presuppose or imply that moral dilemmas are in principle resolvable. This might be taken as supporting evidence for the contrapositive view that the existence of irresolvable moral conflicts leads to some form of relativism. Immanuel Kant said . . . More recently Kurt Baier endorsed a similar position. . . . Ethical egoism and classical act utilitarianism also provide paradigm illustrations of theories precluding irresolvable conflicts. . . . Hence advocates of this position have been called reductionists or systematizers. Hampshire believes the list of historical theories embracing the no-conflicts view is even more comprehensive. He cites Aristotle as a paradigm systematizer and adds Hume, Kant, Mill, Moore, Ross, Prichard, and Rawls. . . . Many of those who adopt the pluralist position on moral dilemmas show substantial sympathy for relativism. Their sense is that absolutist theories according to which there is a single true moral principle or set of principles cannot account for deep irresolvable disagreements. Thus if there are such conflicts, absolutism is false and presumably some form of relativism is true. . . . Replies that have been made to these arguments against irresoluble conflicts, however, do not presuppose or require relativism. One can, as Lemmon did, argue against the "ought" implies "can" principle, without being committed to the view that ought statements are relative to a culture or set of standards. Similarly, there can be relativized and nonrelativized versions of deontic axioms and principles; their denial need not imply any obvious sorts of relativism. (Judith Wagner DeCew, "Moral Conflicts and Ethical Relativism," Ethics 101 (1990): 27–41.

A similar proliferation of concepts and positions can be identified in contemporary work in the philosophy of science.

5. See From Max Weber, ed. H. H. Gerth and C. Wright Mills (New York: Oxford University Press, 1946), esp. pp. 59–60, 294, 323–24; Clifford Geertz, The Interpretation of Cultures (New York: Basic Books, 1973). On the role of intuition in Kant's epistemology and the extent to which the Kantian understanding reflects modernity more generally, see Lachterman, Ethics of Geometry: A Genealogy of Modernity (New York: Routledge, 1989), pp. 9–24.

6. For an illustration of the use of quantum mechanics as the foundation of postmodernism, see Gus diZerega, "Integrating Quantum Theory with Post-Modern Political Thought and Action: The Priority of Relationships over Objects," in Quantum Politics ed. Theodore L. Becker (New York: Praeger, 1991), pp. 65–97.

7. In his recent summary of the arguments against relativism in contemporary philosophy of science, Larry Laudan focuses on three main approaches: positivism (the approach of the Hempel, Oppenheim, Carnap, Reichenbach, and the Vienna Circle), realism (the "falsificationist" view of Sir Karl Popper and others), and pragmatism (his own position, derived from Pierce and Dewey). See Laudan, Science and Relativism, a helpful reminder that the extreme relativist view of the scientific process is probably a minority view among philosophers of science, not the con-

sensual view as many students of literature would like to believe. For the realist position, see Richard W. Miller, *Fact and Method: Explanation, Confirmation and Reality in the Natural and Social Sciences* (Princeton, N.J.: Princeton University Press, 1987).

8. See Hempel, "The Function of General Laws in History," *Journal of Philosophy*, 39 (1942): 35–48 (cited in chap. 3 above). This view was described as the "covering law" model of science because it claims that a scientific explanation takes the form of a general law from which the specific events observed can be logically deduced; hence, the general law can be said to cover the specific case. See also Carl Hempel, *Aspects of Scientific Explanation* (New York: Macmillan, 1965); Carl G. Hempel and Paul Oppenheim, "The Logic of Explanation," *Philosophy of Science* 15 (1948): 135–74.

9. Karl Popper, *Logic of Scientific Discovery* (London: Hutchinson, 1959).

10. For an excellent discussion of falsification and its limitations as the *sole* criterion of scientific knowledge, see Kitcher, *Vaulting Ambition* (Cambridge, Mass.: MIT Press, 1985), pp. 58–72. As Kitcher shows, theoretical clarity is also required if one is to use evidence to assess the truth of a proposition or prediction. Although Kitcher's discussion concerns Darwinian evolutionary theory, his careful analysis has more general relevance.

11. Richard Lewontin, Steven Rose, and Leon Kamin, *Not in Our Genes* (Boston: Pantheon, 1984), p. 32*n*. This example is particularly relevant because it appears in the context of a vigorous critique of the ideological uses of contemporary biology. Since Lewontin, Rose, and Kamin properly distinguish between the truth of a scientific proposition and its political or ideological uses, their definition is particularly appropriate to the discussion here.

12. On the theoretical problem of defining a species, see Ernst Mayr, *Animal Species and Evolution* (Cambridge: Harvard University Press, 1963). The "reality" of the resulting distinctions is suggested by the finding that, in general, people of preliterate, so-called primitive cultures generally identify the same species as do modern biologists (Claude Lévi-Strauss, *La pensée sauvage* [Paris: Plon, 1962]). To be sure, higher-order categorization of animals (e.g., into mammals, reptiles, etc.) seems to be a cultural artifact, as was emphasized by Michel Foucault and other postmodernists (see n. 15 to this chap.). Hence, although most human societies distinguish between animals according to biological standards of taxonomy, the observational constraints on pattern matching do not in themselves resolve the problems of theoretically significant classification.

13. For the observational techniques involved, see any good guide to birds; e.g., National Geographic Society, *Field Guide to the Birds of North America*, 2nd ed. (Washington, D.C.: National Geographic Society, 1987), pp. 8–14.

14. See Howard Margolis, *Patterns, Thinking and Cognition: A Theory of Judgment* (Chicago: University of Chicago Press, 1987). Compare also Henry Pierce Stapp, "A Quantum Theory of the Mind-Brain Interface," Lawrence Berkeley Laboratory Report LBL-28574 Expanded (1990); Roger Penrose, *The Emperor's New Mind* (New York: Oxford University Press, 1989). On pattern matching in facial recognition by the human brain, see n. 37, this chapter, and the discussion of neuroscience in "Religious Faith and Reason," chap. 9, below.

15. A group of French scholars have been among the most influential spokesmen for this perspective; e.g., Jacques Lacan, *Ecrits* (Paris: Le Seuil, 1966); Jacques Derrida, *De la grammatologie* (Paris: Editions de Minuit, 1967); Michel Foucault,

Les mots et les choses (Paris: Gallimard, 1966), translated as *The Order of Things* (New York: Viking, 1970). Similar views, which can be traced back to Nietzsche, were expressed a half-century ago by Walter Lippmann, *A Preface to Morals* (New York: Macmillan, 1929). For a genealogy of this critique, see Allan Bloom, *The Closing of the American Mind* (New York: Simon and Schuster, 1986), esp. pt. 2.

16. Among philosophers of science taking this view, see William V. Quine, *Ontological Relativity and Other Essays* (New York: Columbia University Press, 1969); David Bloor, *Knowledge and Social Imagery* (London: Routledge, 1976); Imre Lakatos, *The Methodology of Scientific Research Programmes* (Cambridge: Cambridge University Press, 1978).

17. For an analysis of the ideological character of modern science that explicitly rejects intuitionism and properly focuses on the distinction between science and its uses, see Lewontin, Rose, and Kamin, *Not in Our Genes*, esp. chap. 3. For the argument that the extreme relativists either rely on intuition or are blatantly self-contradictory in their epistemology, see Laudan, *Science and Relativism*, esp. pp. 146–69. The notes citing the relativist school throughout Laudan's book provide numerous examples of the assertion that science ultimately relies on "belief" rather than on knowledge as conventionally understood by most philosophers and scientists.

18. Richard Rorty, *Contingency, Irony and Solidarity* (Cambridge: Cambridge University Press, 1989). On the assumption that it is no longer possible to discover the true nature of society and morality, Rorty proposes the development of a vocabulary "around notions of metaphor and self-creation" (p. 44). Intellectual progress comes from "an improved self-description rather than a set of foundations" (p. 52). "To see one's language, one's conscience, one's morality, and one's highest hopes as contingent products . . . is to adopt a self-identity which suits one for citizenship in such an ideally liberal state" (p. 61). I am indebted to Suzanne Dovi, whose unpublished paper on Rorty, Mill, and Nietzsche suggested to me the importance of these formulations and their reliance on a distinction between "public" and "private" life, which Rorty himself does not view as contingent or problematic.

19. At the extreme, this skepticism is tantamount to a return to Descartes' thought experiment of imagining an evil genius who deceived humans about every element of their existence. See René Descartes, *Discourse on Method*, trans. Paul J. Oscamp (New York: Bobbs-Merrill, 1965). Descartes' answer to this challenge—the deduction of his own existence from the fact of thought ("I think, therefore I am")—has been challenged on the grounds that it is embedded in, and presupposes, a linguistic and cultural tradition in which the first person singular ("I") is the appropriate locus for propositions about the existence or nonexistence of a being. On the continued relevance of the Cartesian thought experiment—and the extent to which modern science provides startling answers to it, challenging the conventional mind/body dualism—see Daniel C. Dennett, *Consciousness Explained* (Boston: Little, Brown, 1991).

20. Lewontin, Rose, and Kamin, *Not in Our Genes*, p. 57. See also Leon Kamin, *The Science and Politics of I.Q.* (Potomac, Md.: Lawrence Erlbaum, 1974).

21. Kitcher's analysis of "pop sociobiology" provides an excellent illustration of how it is possible and necessary to go beyond mere hunches when evaluating scientific ideas. As he points out, for example, "much of the sociobiological literature is marked by a general approach to human behavior that can easily be used

to support harmful views. Because they stress the genetic basis of behavior, many sociobiologists *seem* to be endorsing a strategy of linking behavioral differences to genetic differences, and this strategy encourages the denigrating of particular racial and social groups" (*Vaulting Ambition*, p. 6). The problem, of course, is twofold. First, what is "harmful"? Second, why does the intellectual "strategy" have these consequences? In other contexts, evidence "linking behavioral differences to genetic differences" has been used to explain diversity in individual learning styles and performance and hence to encourage educational reforms that have the opposite effect. See Howard Gardner, *Frames of Mind* (New York: Basic Books, 1983). Intuition, therefore, is not enough. As Kitcher puts it, the issue is "*a dispute about evidence*" (*Vaulting Ambition*, p. 8).

22. A. Michael Warnecke, Roger D. Masters, and Guido Kempter, "The Roots of Nationalism: Nonverbal Behavior and Xenophobia," *Ethology and Sociobiology* 13 (1992): 267–82; A. Michael Warnecke, *The Personalization of Politics: An Analysis of Emotion, Cognition, and Nonverbal Cues* (Senior Fellowship Thesis, Dartmouth College, 1991). A follow-up experiment has confirmed these negative reactions to foreign leaders and reveals that the terms used to describe unknown leaders tend to reflect national stereotypes even though the viewer is unaware of each individual's country of origin. Speaking broadly, the German leaders are described as "ugly" and "boring"; the French as "energetic" and "cheerful"; and the Americans as "powerful" and "competent"—even though their nationality was never identified (Siegfried Frey, Roger D. Masters, and Guido Kempter, unpublished data). See also Edward T. Hall, *The Silent Language* (Greenwich, Conn.: Fawcett, 1959).

23. For a review, see Pawel Lewicki, Thomas Hill, and Maria Czyzewska, "Nonconscious Acquisition of Information," *American Psychologist* 47 (1992): 796–801.

24. To cite but one recent example, "the view of coevolution" of parasites and host animals that "makes intuitive sense" has been rejected by biologists as "misguided" because "virtually none of it is consistent with deeper, more rigorous evolutionary thinking" (John Rennie, "Living Together," *Scientific American* 266 [1992]: 127, citing Douglas E. Gill of the University of Maryland).

25. Note that, in Figure 4.1, the cognitive assessments of French leaders were more favorable than those of Germans but that this difference did not occur in either emotional responses while watching the video excerpts or in the overall ratings on the 0–100 thermometer scale (known to predict, for example, voters' choices during elections).

26. Laudan, *Science and Relativism*, p. 80.

27. *Notebooks of Leonardo da Vinci*, ed. I. Richter (Oxford: Oxford University Press, 1980), p. 8.

28. For the logic and mathematics, see R. A. Fisher, "The Mathematics of a Lady Tasting Tea," in his *Design of Experiments* (Adelaide: Hafner Press, Macmillan, 1971). In this regard, it is interesting to note the role of technology in changes in scientific theory and method. The computer has greatly increased the visibility of these mathematical measures of the "significance" (or probability) of outcomes and therewith the reliability of the inferences made concerning them. A generation ago, the computation of statistical measures of correlation and significance was laborious: I can recall devoting an entire day to the computation of a single multiple regression equation. Today I can compute the same statistic in a minute or less on

my personal computer. The result, as anyone familiar with empirical work in social psychology can attest, is a proliferation of statistical measures (often imperfectly understood by the scientist using them).

29. In this case, the "bias" seems to *strengthen* the finding that Americans respond negatively to silent images of foreigners rather than to undermine it. Personality was measured by using a questionnaire developed by C. Robert Cloninger, who has argued that individual character traits vary on three dimensions: harm avoidance (fear vs. insensitivity to danger and risk), reward dependence (outer-directed vs. inner-directed behavior), and novelty seeking (boredom with routine vs. need for novelty). See C. Robert Cloninger, "A Unified Biosocial Theory of Personality and Its Role in the Development of Anxiety States," *Psychiatric Developments* 3 (1986): 167–226; and "A Systematic Method of Clinical Description and Classification of Personality Variants," *Archives of General Psychiatry* 44 (1987): 573–88. The table below compares national averages on each of the three dimensions for males and females with the personality traits of the subjects in the experiment described in n. 22 above (Warnecke, Masters, and Kempter, "The Roots of Nationalism") and a second, follow-up experiment:

	National Sample	Warnecke Experiment	Frey–Masters Experiment
Novelty seeking			
Males	13.5	14.9	18.9
Females	13.1	17.3	19.9
Average	13.3	16.1	19.5
Harm avoidance			
Males	10.7	10.4	10.0
Females	12.8	13.2	11.2
Average	11.7	11.8	11.1
Reward dependence			
Males	19.8	20.0	19.1 **
Females	21.4	22.2	24.7 **
Average	20.6	21.1	22.2

**Significant gender difference between males and females ($p < .003$) within experimental sample

It seems that the procedures used to recruit subjects discouraged individuals with a low tolerance for novelty. In the current case, one can presume that such individuals would be even more likely to respond negatively to strangers, but in general it is impossible to know whether this is the case unless the confounding factors have been isolated and measured.

30. This assumption of modern scientific methods is in tension with the existence of individual differences in mathematics or other specific cognitive abilities: I have never been able to learn the calculus and I find abstract mathematical problems impenetrable, just as I cannot carry a tune. On the character and biological origins of such individual differences in what is conventionally called intelligence (supposedly measured by IQ tests), see Howard Gardner, *Frames of Mind*. Even granting these differences, however, modern scientists could reply that they do not coincide with the social institutions that other cultures developed to ensure that only the "elect" had access to ritual or privileged knowledge. From the verification-ist point of view, equal opportunity is required because no other selection process

focuses entirely on cognitive ability to the exclusion of socioeconomic, religious, or ethnic categories that are ascriptive rather than achievement-oriented. Nonetheless, the practice of IQ testing and egalitarian rhetoric can readily lead to the absurdity of nonscientists claiming to reject scientific evidence as untrue merely because it contradicts their opinions, prejudices, or intuitions.

31. See Foucault, *The Order of Things*. This insight can be traced to Marx's critique of the "fetishism of commodities" in the *Economic and Philosophic Manuscripts of 1844*, ed. Dirk J. Struik (New York: International Publishers, 1964).

32. Is modern science a mode of knowing that reflects a particular political system and its unstated social values? Can all knowledge be limited by the cultural context in which it is discovered and formulated? This criticism has typically been rejected on the grounds that the circumstances for discovering the truth are not the same thing as the truth itself. The classic statement of this point in the philosophy of science is Hans Reichenbach, *Experience and Prediction* (Chicago: University of Chicago Press, 1938). See also Miller, *Fact and Reason*.

33. Compared to Lachterman's careful analysis of ancient geometry (esp. *Ethics of Geometry*, chap. 2), the presentation here will doubtless seem highly superficial to specialists. See also D. H. Fowler, *The Mathematics of Plato's Academy* (Oxford: Clarendon Press, 1987).

34. For a typical caricature of Plato's theory of knowledge, see T. D. Weldon, *The Vocabulary of Politics* (New York: Penguin, 1973).

35. This difference is particularly evident in the thinkers like Heraclitus, Aristotle, and Lucretius, for whom material and efficient causes—the center of modern concerns—were explicitly introduced as legitimate concerns for science. For Heraclitus, the visible things are mere flux or change, but behind them is a cosmos based on formal principles; in the case of Aristotle, formal and final causes are primary; for Lucretius, thought about the "nature of things" has the end or purpose of justifying the philosophic life. Even when ancient thinkers deny the primacy of form in the visible world, therefore, the question of the shapes of things and its related concern for the best or most excellent form of a thing, remains primary. Cf. Philip Wheelwright, *The Presocratics* (New York: Odyssey, 1966); Arlene W. Saxonhouse, *Fear of Diversity: The Birth of Political Science in Ancient Greek Thought* (Chicago: University of Chicago Press, 1992); Leo Strauss, *The City and Man* (Chicago: Rand McNally, 1964); Sheldon Wolin, *Politics and Vision* (London: Allen Unwin, 1961); Irving M. Zeitlin, *Plato's Vision* (Englewood Cliffs, N.J.: Prentice-Hall, 1992).

36. While the overly simplified right brain/left brain dichotomy is almost surely incorrect, some major differences are well founded. For example, the formation of mental images—popularly attributed entirely to the right hemisphere—involves some neuronal processing on both sides of the brain (see Steven M. Kosslyn, "Aspects of Cognitive Neuroscience of Mental Imagery," *Science* 240 [1988]: 1621–26), but some differences in typical hemispheric function—such as the human species' tendency to right-handedness as a result of the specialization of the left hemisphere in the control of fine motor coordinations—are difficult to deny. See Norman Geschwind and Albert M. Galaburda, *Cerebral Lateralization: Biological Mechanisms, Associations, and Pathology* (Cambridge, Mass.: MIT Press, 1987).

37. Malcolm P. Young and Shigero Yamane, "Sparse Population Coding of Faces in the Inferotemporal Cortex," *Science* 256 (1992): 1327–31; Edmund Rolls, "Information Representation, Processing and Storage in the Brain: Analysis at the Single Neuron Level," in *The Neural and Molecular Bases of Learning*, ed. J. P.

Changeux and M. Konishi (New York: Wiley & Sons, 1987); Edmund Rolls, "The Representation and Storage of Information on Neuronal Networks in the Primate Cerebral Cortex and Hippocampus," in *The Computing Neuron*, ed. R. Burbin, C. Miall, and G. Mitchison (New York: Addison Wesley, 1989); and Edmund Rolls, "The Processing of Face Information in the Primate Temporal Lobe," in *Processing Images of Faces*, ed. V. Bruce and M. Burton, (London: Ablex, 1989).

38. For the processing of verbal language, see Steven E. Peterson, Peter T. Fox, Michael I. Posner, Mark Mintun, and Marcus E. Raichle, "Positron Emission Tomographic Studies of the Processing of Single Words," *Journal of Cognitive Neuroscience* 1 (1989): 153–70; Steven E. Petersen, Peter T. Fox, Abraham Z. Snyder, and Marcus E. Raichle, "Activation of Extrastriate and Frontal Cortical Areas by Visual Words and Word-Like Stimuli," *Science* 249 (1990): 1041–44. The practical implications of these findings can be illustrated by Descartes's argument that, "because the concept of figure is so common and simple that it is involved in every object of sense," it is useful to abstract "from every other feature" and focus merely on its mathematical or geometrical representation. "Thus whatever you suppose colour to be, you cannot deny that it is extended and in consequence possessed of figure. Is there then any disadvantage if, while taking care not to admit any new entity uselessly, or rashly to imagine that it exists, and not denying indeed the beliefs of others concerning colour, but merely abstracting from every other feature except that it possesses the nature of figure, we conceive the diversity existing between white, blue, and red, etc., as being like the difference between the following similar figures?" (*Rules for the Direction of the Mind*, Rule 12.) The scientific measurement of colors by their wavelength is, of course, precisely what Descartes had in mind (I am indebted to Leon Kass for this example, which played a central role in his lecture to the Claremont Institute Conference on the "Permanent Limitations of Modern Science"). Although modern scientists in many fields would agree with Descartes's procedure, there *is* a major disadvantage to such a representation when studying the way the brain actually sees color. Continuous mathematical measures such as the wavelength of light obscure the way the structural organization of the cortex reflects functional adaptations to "natural" features of reality.

39. In a review of a volume surveying research on the visual system, Terrence J. Sejnowski summarizes the contemporary neuroscientific approach as follows:

Models simulated on digital computers run the risk of being too accurate. For example, the array of pixels in a digital image is perfectly ordered, but in the retina individual photoreceptors form an irregular lattice. . . . Rectifying nonlinearities are found already in simple cells, and complex cells are even more nonlinear . . . cortical processing is essentially nonlinear within its normal operating range. . . . Under natural conditions, the shape of an object can be sensed by using several cues . . . such as stereo, texture, and shading . . . images of the natural world have a self-similar structure. Thus the world is not arbitrary, and this regularity is probably exploited by the brain in its representation of visual information. It is well established by physiological recordings that many cues, such as disparity, are represented in the visual cortex by populations of neurons that have broad, overlapping selectivities. Recent evidence also points toward a sparse population code for complex objects such as faces. . . . We need to develop a better understanding of the types of distributed representations found in the cerebral cortex. (Review of *Computational Models of Visual Processing*, ed. Michael S. Landy and J. Anthony Movshon [Cambridge, Mass.: MIT Press, 1991], in *Science* 257 [1992]: 687–88).

The disadvantage of Descartes's procedure is that, used alone, one would confuse the digital computer with the human brain, failing to understand how we perceive the world (and therewith taking on faith the assumption that there is no natural congruence between some aspects of the reality of color and our perceptions of it).

40. Howard Margolis, *Patterns, Thinking, and Judgment* (Chicago: University of Chicago Press, 1987).

41. The parallel between the three modes of knowing described in chapter 1 (depicted in Figure 1.1) and this discussion will not have escaped the thoughtful reader. But it is of immense importance that intuition, verification, and pattern matching are modes of discovering knowledge *within* the community of contemporary, secular scientists.

42. See Plato's *Apology of Socrates*, and compare Xenophon, *Apology of Socrates*; Aristotle, *Metaphysics*, 13.1078b15–31 and *Sophistical Refutations*, 34.183b1–15; and Plutarch, *Adversus Colotem*, 1118C.

43. Aristotle, *Parts of Animals*, 1.642a25–31. Cf. *Eudemean Ethics*, 1.5.1216b1–25; *On Generation and Corruption*, 1.2.315a26–315b6.

44. Unlike earlier teachers and schools, both the Lyceum and the Academy provided the models for the self-perpetuating institutions of teaching and research that came to be the Western university. That Aristotle should be considered a Socratic is emphatically stated by Cicero, one of the rare thinkers to have read Aristotle's works before they were apparently combined with manuscripts of Theophrastus by Andronicus of Rhodes. See Roger D. Masters, "The Case of Aristotle's Missing Dialogues," *Political Theory* 5 (1977): 31–60, and "On Chroust: A Reply," *Political Theory* 7 (1979): 545–47.

5. The Nature of Social Communication

1. See the selections by Hempel, Nagel, and others in May Broadbeck, ed., *Readings in the Philosophy of the Social Sciences* (New York: Macmillan, 1968); or in Leonard I. Krimerman, ed., *The Nature and Scope of Social Science: A Critical Anthology* (New York: Appleton Century Crofts, 1969). A frequently used example is the deduction that a bird is a swan from the predicate of its whiteness (a syllogism of the form: "All swans are white; X is white; therefore, X is a swan"). I recall a lecture in which Thomas Kuhn, the eminent historian of science, used this example; in conversation after the lecture, Kuhn expressed surprise that the question of how a swan recognized another swan was an empirical problem that had been explored by ethologists like Konrad Lorenz for over a generation. See the citations in n. 3, this chapter.

2. For a precedent, see Locke's discussion of the duty of parents to care for their children in *Essay concerning Human Understanding*, analyzed below in chap. 7.

3. For what is perhaps the classic statement, see Konrad Z. Lorenz, *Studies in Animal and Human Behavior* (Cambridge, Mass.: Harvard University Press, 1970–71), vols. 1 and 2; and for a more popular introduction to Lorenz's view of animal behavior, *On Aggression* (New York: Harcourt Brace Jovanovich, 1966). See also Robert Hinde, *Ethology* (Glasgow: William Collins Sons, 1982); Mario von Cranach, Klaus Foppa, Wolfgang Lepenies, and Detlev Ploog, eds. *Human Ethology: Claims and Limits of a New Discipline* (Cambridge: Cambridge University Press, 1979).

4. Charles Darwin, *The Expression of Emotion in Animals and in Man* (Chicago: University of Chicago Press, 1962); Jan A.R.A.M. Van Hooff, "The Facial Displays of the Catyrrhine Monkeys and Apes," in *Primate Ethology*, ed. Desmond Morris (New York: Anchor Books, 1969), and "A Comparative Approach to the Phylogeny of Laughter and Smiling," in *Non-verbal Communication*, ed. Robert A. Hinde (New York: Cambridge University Press, 1972), pp. 209–41; Paul Ekman and Harriet Oster, "Facial Expressions of Emotion," *Annual Review of Psychology* 30 (1979):527–54.

5. While the demonstration of intentionality in nonhuman animals is exceptionally difficult, there are now well-documented instances of the deceptive use of nonverbal displays by chimpanzees, vervet monkeys, and other primates; in these instances, some individuals have been shown to use a signal "incorrectly" as means of manipulating others. See Roger D. Masters, "Conclusion," in *Primate Politics*, ed. Glendon Schubert and Roger D. Masters (Carbondale: Southern Illinois University Press, 1991).

6. James Barham, "A Poincaréan Approach to Evolutionary Epistemology", *Journal of Social and Biological Structures* 13 (1991):193–258; James H. Fetzer, *Sociobiology and Epistemology* (Dordrecht: Reidel, 1985). The extraordinarily rich analyses of animal behavior over the past generation, including the magnificent long-term field studies of nonhuman primates, have contradicted the traditional distinction between humans and animals and made it necessary to reconsider epistemological issues in the light of our evolutionary past. See Vicki Hearne, *Adam's Task* (New York: Viking, 1987), and for exceptionally fine popularizations of extended field research among nonhuman primates, Shirley C. Strum, *Almost Human: A Journey into the World of Baboons* (New York: W. W. Norton, 1987); Jane Goodall, *Through a Window* (Boston: Houghton Mifflin, 1990). It may be no accident that the authors of the last three works cited are women; see Glendon Schubert, "Introduction," in *Primate Politics*, ed. Glendon Schubert and Roger D. Masters (Carbondale: Southern Illinois University Press, 1991).

7. Robin Fox, ed., *Biosocial Anthropology* (London: Malaby, 1975); Irenäeus Eibl-Eibesfeldt, *Ethology: The Biology of Behavior*, 2d. ed. (New York: Holt, Rinehart & Winston, 1975), and *Human Ethology* (New York: Aldine de Gruyter, 1989); Michael R. A. Chance, ed., *Social Fabrics of the Mind* (Hillsdale, N.J.: Lawrence Erlbaum, 1989); Glendon Schubert, *Evolutionary Politics* (Carbondale: Southern Illinois University Press, 1989). See also Masters, *The Nature of Politics*; and Glendon Schubert and Roger D. Masters, eds., *Primate Politics*. (Carbondale: Southern Illinois University Press, 1991).

8. Social structures in primates vary from asociality (e.g., orangutans) to groups with linear dominance hierarchies (e.g., rhesus monkeys) or troops in which an adult male herds several females and their young (e.g., hamadryas baboon) (Irven DeVore, *Primate Behavior* [New York: Holt, Rinehart, and Winston, 1965]; Richard Wrangham, "On the Evolution of Ape Social Systems," *Biology and Social Life* 18 [1979]:335–68).

9. On the discipline of behavioral ecology as a way of explaining differences in primate social structure, see David Barash, *Sociobiology and Behavior*, 2nd ed. (New York: Elsevier, 1982); Eric Alden Smith and Bruce Winterhalder, *Evolutionary Ecology and Human Behavior* (New York: Aldine de Gruyter, 1992); Robin I. M. Dunbar, *Primate Social Systems* (Ithaca, N.Y.: Cornell University Press, 1988), and chap. 7 below.

10. See J.A.R.A.M. van Hooff, "A Structural Analysis of the Social Behavior of a Semi-captive Group of Chimpanzees," in *Social Communication and Movement*, ed. Mario von Cranach and Ian Vine (New York: Academic Press, 1973); Robert Plutchik, *Emotion: A Psychoevolutionary Synthesis* (New York: Harper and Row, 1980); and Roger D. Masters, Denis G. Sullivan, John T. Lanzetta, Gregory J. McHugo, and Basil G. Englis, "The Facial Displays of Leaders: Toward an Ethology of Human Politics," *Journal of Social and Biological Structures* 9 (1986): 319–43.

11. Hanus Papousek and Mechthild Papousek, "Early Ontogeny of Human Social Interaction: Its Biological Roots and Social Dimensions," in *Human Ethology*, ed. Mario von Cranach et al., (Cambridge: Cambridge University Press, 1979), pp. 456–78.

12. Hans Hass, *The Human Animal* (London: Hodder and Stoughton, 1970).

13. Carroll E. Izard, Elizabeth A. Hembree, and Robin R. Huebner, "Infants' Emotion Expressions to Acute Pain: Developmental Change and Stability of Individual Differences," *Developmental Psychology* 23 (1987): 105–13.

14. Hubert Montagner, *L'Enfant et la communication* (Paris: Stock, 1978) and, for visual documentation, "Silent Speech," in "Horizon" (BBC2 Videotape, July 28, 1977). See also the photographs in Eibl-Eibesfeldt's *Human Ethology*.

15. Gail Zivin, "Facial Gestures Predict Preschooler's Encounter Outcomes," *Social Science Information* 16 (1977):715–29; Hubert Montagner, A. Restoin, D. Rodriguez, and F. Kontar, "Aspects fonctionels et ontogénétiques des interactions de l'enfant avec ses pairs au cours des trois premières années," *Psychiatrie de l'enfant* 31 (1988): 173–278.

16. Fred Strayer, "The Organization and Coordination of Asymmetrical Relations among Young Children: A Biological View of Social Power," in *Biopolitics: Ethological and Physiological Approaches*, ed. Meredith Watts, New Directions for Methodology of Social and Behavioral Science, no. 7 (San Francisco: Jossey-Bass, 1981), pp. 33–50; William Charlesworth, "The Child's Development of the Sense of Justice: Moral Development, Resources, and Emotions," in *The Sense of Justice: Biological Foundations of Law*, ed. Roger D. Masters and Margaret Gruter (Newbury Park, Calif: Sage, 1992), pp. 256–77. On the analogous role of reassurance and bonding in nonhuman primates who were once viewed as relying primarily on agonic behavior for dispute settlement, see Frans B. M. deWaal, *Peacemaking among Primates* (Cambridge, Mass.: Harvard University Press, 1989).

17. Carol Barner-Barry, "Longitudinal Observational Research and the Study of the Basic Forms of Political Socialization," in *Biopolitics: Ethological and Physiological Approaches*, ed. Meredith Watts (San Francisco: Jossey-Bass, 1981), pp. 51–60. See also the references in the preceding four notes.

18. John T. Lanzetta and Scott P. Orr, "Influence of Facial Expressions on the Classical Conditioning of Fear," *Journal of Personality and Social Psychology* 39 (1980): 1081–87; Scott P. Orr and John T. Lanzetta, "Facial Expressions of Emotion as Conditioned Stimuli for Human Autonomic Responses," *Journal of Personality and Social Psychology* 38 (1980): pp. 278–82.

19. G. E. Schwartz, P. L. Fair, P. Salt, M. R. Mandel, and G. L. Klerman, "Facial Muscle Patterning to Affective Imagery in Depressed and Nondepressed Subjects," *Science* 192 (1976): pp. 489–91.

20. Paul Ekman, R. W. Levenson, and Wallace V. Friesen, "Autonomic Nervous System Activity Distinguishes among Emotions," *Science* 221 (1983): 1208–10.

21. Paul Ekman, Wallace V. Friesen, and Phoebe Ellsworth, *Emotion in the Human Face* (New York: Pergamon, 1972); Paul Ekman, Wallace V. Friesen, and S. Ancoli, "Facial Signs of Emotional Expression," *Journal of Personality and Social Psychology* 39 (1980): 1125–34.

22. Roger D. Masters and Jean Mouchon, "Les gestes et la vie politique," *Le français dans le monde*, no. 203 (1986):85–87; Roger D. Masters and Denis G. Sullivan, "Nonverbal Displays and Political Leadership in France and the United States," *Political Behavior* 11 (1989): 121–53. For more details and an explanation of why I have cited my own work at such length, see *The Nature of Politics*, chap. 2, and the "Facial Displays and Political Leadership" section of this chapter.

23. Frans de Waal, *Chimpanzee Politics* (Cambridge, Mass.: Harvard University Press, 1982); Jane Goodall, *The Chimpanzees of Gombe* (Cambridge, Mass.: Harvard University Press, 1986); Roger D. Masters, Denis G. Sullivan, Alice Feola, and Gregory J. McHugo, "Television Coverage of Candidates' Display Behavior during the 1984 Democratic Primaries in the United States," *International Political Science Review* 8 (1987): pp. 121–30; Masters et al., "The Facial Displays of Leaders."

24. de Waal, *Peacekeeping among Primates*; Chance, *Social Fabrics of the Mind*; Denis G. Sullivan and Roger D. Masters, "Biopolitics, the Media and Leadership: Nonverbal Cues, Emotions and Trait Attributions in the Evaluation of Leaders," in *Research in Biopolitics*, ed. Albert Somit and Steven Peterson, Greenwich, Conn.: JAI Press, in press).

25. For example, during the French legislative election campaign in 1985, there was a televised debate between then Prime Minister Laurant Fabius (Socialist) and his principal challenger, Jacques Chirac (Gaullist mayor of Paris); Fabius engaged in an unprovoked aggressive display toward Chirac, which seemingly produced a sharp reversal in their popularity: see *Figaro Magazine*, November 9, 1985, pp. 126–27.

26. Daniel Tranel and Antonio R. Damasio, "Learning without Awareness: An Autonomic Index of Facial Recognition by Prosopagnosics," *Science* 218 (1985): 1453–1454; Daniel Tranel, Antonio R. Damasio, and Hanna Damasio, "Intact Recognition of Facial Expression, Gender, and Age in Patients with Impaired Recognition of Face Identity," *Neurology* 38 (1988):690–96.

27. Masters, *The Nature of Politics*, p. 254.

28. See Young and Yamane, "Sparse Population Coding of Faces in the Inferotemporal Cortex," and Edmund T. Rolls, "The Processing of Face Information in the Primate Temporal Lobe," in *Processing Images of Faces*, ed. V. Bruce and M. Burton (London: Ablex, 1989).

29. Arthur Kling, "Neurological Correlates of Social Behavior," in *Ostracism: A Social and Biological Phenomenon*, ed. Margaret Gruter and Roger D. Masters (New York: Elsevier, 1986), pp. 27–38; Arthur Kling, "Brain Mechanisms and Social/Affective Behavior," *Social Science Information* 26 (1987): 375–84.

30. E.g., Bernard Mandeville, *The Fable of the Bees* (Oxford: Clarendon Press, 1924), vol. 2, pp. 286–87; Rousseau, *Second Discourse*, ed. Roger D. Masters (New York; St. Martin's Press, 1964), pp. 122–23.

31. Robert Frank, *Passions within Reason* (New York: Norton, 1989); Kevin B. MacDonald, *Social and Personality Development: An Evolutionary Synthesis* (New York: Plenum, 1989).

32. Richard Peters and Henri Taijfel, "Hobbes and Hull: Metaphysicians of Be-

havior," in *Hobbes and Rousseau*, ed. Maurice Cranston and Richard S. Peters (New York: Doubleday Anchor, 1972), pp. 165–83; James Q. Wilson, Presidential Address to American Political Science Association, Chicago, September 4, 1992. For the theoretical importance of these assumptions, see Masters, *The Nature of Politics*, chaps. 1, 2; Schubert and Masters, *Primate Politics*, esp. chap. 10.

33. The tabula rasa view of human nature, which characterized not only orthodox behaviorism but modern social theories more generally, treats the central nervous system as a "black box" whose responses are primarily if not entirely shaped by experience. Such a model assumes that the behavioral effects of stimuli depend on the contingencies of reinforcement to which the organism has been exposed. Since it was not possible to look inside the black box and the postulation of invisible and undemonstrable mental phenomena was rejected as unscientific, psychologists like Watson, Hull, and Skinner sought "laws of behavior" based on predictable relations between stimuli and response; e.g., B. F. Skinner, *Science and Human Behavior* (New York: Free Press, 1965). Because behaviorists focused on such extrinsic factors as rates of reinforcement, which apply in similar ways to different species, they rarely asked why a particular environmental cue is a stimulus for a given organism; in contrast, the tradition of ethology epitomized by Lorenz's studies of "innate releasing mechanism" gives primacy to heritable behaviors that can evolve a communicative function: see Konrad Z. Lorenz, *Evolution et modification du comportement* (Paris: Payot, 1967) or Konrad Z. Lorenz and Paul Leyhausen, *Motivation of Human and Animal Behavior* (New York: Van Nostrand Reinhold, 1973). It is ironic that Marxists share the environmentalist premises of Anglo-Saxon psychologists who were, in politics, resolutely hostile to communism; cf. Richard Lewontin, Steven Rose, and Leon Kamin, *Not in Our Genes* (New York: Pantheon, 1984).

34. Edward Dolnick, "What Dreams Are (Really) Made Of," *Atlantic* 266 (1990):41–61.

35. E.g., José Ambrose-Ingerson, Richard Granger, and Gary Lynch, "Simulation of Paleocortex Performs Hierarchical Clustering," *Science* 247 (1990): 1344–48. *Note:* In this section of my argument, citations will represent a somewhat arbitrary sampling of an enormous and rapidly growing literature; they give the reader a sense of the research with which I am familiar, not a complete survey of the state of the field (which would in any event soon be obsolete). For recent surveys, see Michael Gazzaniga, *The Social Brain* (New York: Basic Books, 1985); William Allman, *Apprentices of Wonder* (New York: Bantam, 1989); Stephen M. Kosslyn and Olivier Koenig, *Wet Mind: The New Cognitive Neuroscience* (New York: Free Press, 1992); *Scientific American*, "Mind and Brain" (Special Issue), 267 (September 1992): 48–159.

36. Maurizio Corbetta, Francis M. Miezin, Susan Dobmeyer, Gordon L. Shulman, and Steven E. Peterson, "Attentional Modulation of Neural Processing of Shape, Color, and Velocity in Humans," *Science* 248 (1990): 1556–59.

37. Gerald M. Edelman, *Topobiology* (New York: Basic Books, 1989).

38. George Shepherd, *Neurobiology* (New York: Wiley, 1985); Masters, *The Nature of Politics*, Figures 2.4, 3.3.

39. For visual images of the diverse sites in the brain that are active in varied types of reading task, see Steven E. Peterson, Peter T. Fox, Michael I. Posner, Mark Mintun, and Marcus E. Raichle, "Positron Emission Tomographic Studies of the Processing of Single Words," *Journal of Cognitive Neuroscience* 1 (1989): 153–70;

Steven E. Peterson, Peter T. Fox, Abraham Z. Snyder, and Marcus E. Raichle, "Activation of Extrastriate and Frontal Cortical Areas by Visual Words and Word-Like Stimuli," *Science* 249 (1990): 1041–49.

40. Martha Denckla, "Application of Disconnexion Concepts to Developmental Dyslexia," Geschwind Memorial Lecture, 37th Annual Meeting of the Orton Dyslexia Society, Philadelphia, November 1986. Inglewood, CA: Audio-Stats Educational Services, Tape #916R-34. For a general introduction to the neurological approach to dyslexia, see Richard L. Masland, "Neurological Aspects of Dyslexia," in *Dyslexia Research and Its Applications to Education*, ed. G. T. Pavlidis and T. R. Miles (New York: John Wiley, 1983), pp. 35–66.

41. This hypothesis is strengthened by the similarities between "developmental" deficits (which are perceived only as the child matures) and "acquired" ones (which result from accident or disease, often in adults) (Andrew Ellis, *Reading, Writing and Dyslexia* [Hillsdale, N.J.: Lawrence Erlbaum, 1984]).

42. Benjamin Whorf, *Language, Thought, and Reality* (Cambridge, Mass.: MIT Press, 1956); Roger Brown, *Words and Things* (New York: Free Press, 1958); Eric Lenneberg, ed., *New Directions in the Study of Language* (Cambridge, Mass.: MIT Press, 1964).

43. George Mandler, *Mind and Emotion* (New York: Wiley, 1975); Robert Zajonc, "Feeling and Thinking: Preferences Need No Inferences, *American Psychologist* 35 (1982): 151–75, and "On the Primacy of Affect," *American Psychologist* 39 (1984): 117–23; Richard S. Lazarus, "On the Primacy of Cognition," *American Psychologist* 39 (1984): 124–29.

44. George Marcus, "The Structure of Emotional Appraisal," *American Political Science Review* 92 (1988): 737–62; John L. Sullivan, John H. Aldrich, Eugene Borgida, and Wendy Rahn, "Candidate Appraisal and Human Nature: Man and Superman in the 1984 Election," *Political Psychology* 11 (1990): 459–84.

45. Gazzaniga, *The Social Brain*, p. 75.

46. Ibid. See also Pavel Lewicki, Thomas Hill, and Maria Czyewska, "Nonconscious Acquisition of Information," *American Psychologist* 47 (1992): 796–801.

47. Denis G. Sullivan and Roger D. Masters, " 'Happy Warriors': Leaders' Facial Displays, Viewers' Emotions, and Political Support," *American Journal of Political Science* 32 (1988): 345–68; Roger D. Masters and Denis G. Sullivan, "Nonverbal Displays and Political Leadership in France and the United States," *Political Behavior* 11 (1989): 121–53; A. Michael Warnecke, "The Personalization of Politics" (Senior Fellowship Thesis, Dartmouth College, 1991). In our own experiments, we found that if subjects were not prevented from communicating with others, they routinely "cheated" (i.e., consulted with others) on cognitive questions such as the names of candidates for a presidential election; in contrast, this problem has never arisen with regard to subjects' descriptions and reports of emotional response to leaders. It is little wonder, then, that the Socratic injunction "Know thyself" is the most difficult and essential task of the philosophic life. As a friend put it, "How many humans have self-knowledge or strive for it? And among those who do, how often do we achieve it?"

48. In the technical terms of statistical analysis, the weights and significance of the coefficients in a multiple regression model may be far more appropriate than a zero-order correlation coefficient; in an analysis of variance, interaction terms are often more important than "main effects." Cf. Hume, *Treatise*, 1.3.12, and n. 20, chap. 7, below.

49. Mortimer Mishkin and Thomas Appenzeller, "The Anatomy of Memory," *Scientific American* 256 (1987): pp. 80–89.

50. Peter Brennan, Hideto Kaba, and Eric B. Keverne, "Olfactory Recognition: A Simple Memory System," *Science* 250 (1990): 1223–26.

51. These neuroanatomical details also help to explain some otherwise puzzling features of the response to a leader's nonverbal displays described in detail in "Facial Displays and Political Leadership," below (for references, see esp. n. 92). Psychophysiological studies found that watching excerpts of President Reagan's facial expressions elicited much stronger immediate reactions in viewers' facial muscles and autonomic systems for image-only presentations than for those with sound plus image. Comparing verbally self-reported emotions, responses to sound-only or image-only presentations were not weaker (if anything, stronger) than responses to excerpts seen with both sound and image. Rather than being additive, therefore, it seems that auditory and visual cues are processed separately and that the strongest immediate reactions to facial displays, especially to unambiguous cues of happiness/reassurance, occur in the image-only condition. It is tempting to assume that these findings reflect the fact that image-only excerpts can be processed primarily by the ventral pathway from the visual cortex through the temporal lobes to the limbic system, a pathway that could elicit emotion without activating the rich store of verbal memories and mental images in the neocortex.

This hypothesis might be reinforced by the puzzling way that seeing competing rivals activates prior attitudes in immediate responses to a facial display. When viewers saw Reagan alone, their psychophysical responses were not significantly influenced by their prior attitudes; similarly, when Reagan was the only stimulus figure seen, descriptions of his facial display behavior were not influenced by prior attitudes toward him—except in the sound-plus-image media condition, in which viewers had to integrate a verbal message with the sight of facial displays. But when viewers saw excerpts of both Reagan and Hart in a single study, prior attitudes interacted with display in producing psychophysiological responses. John T. Lanzetta, Denis G. Sullivan, Roger D. Masters, and Gregory J. McHugo, "Viewers' Emotional and Cognitive Responses to Televised Images of Political Leaders," in *Mass Media and Political Thought*, ed. Sidney Kraus and Richard Perloff (Beverly Hills, Calif.: Sage, 1985), pp. 85–116; Gregory J. McHugo, John T. Lanzetta, Denis G. Sullivan, Roger D. Masters, and Basil G. Englis, "Emotional Reactions to Expressive Displays of a Political Leader," *Journal of Personality and Social Psychology* 49 (1985): 1512–29; Gregory J. McHugo, John T. Lanzetta, and Lauren K. Bush, "The Effect of Attitudes on Emotional Reactions to Expressive Displays of Political Leaders, *Journal of Nonverbal Behavior* 15 (1991): 19–41. Similarly, competitive situations, especially in France, produced strong interactions between prior attitudes and the descriptions of a leader's nonverbal behavior. These apparently confusing findings illustrate how context can change the way prior cognitions interact with emotions.

52. See Mishkin and Appenzeller, "The Anatomy of Memory"; Larry Squire, "The Mechanisms of Memory, *Science* 232 (1986): 1612–19; Gregory J. McHugo, C. A. Smith, and John T. Lanzetta, "The Structure of Self-reports of Emotional Response to Film Segments," *Motivation and Emotion* 6 (1982): 365–85.

53. Larry Squire, "Brain and Memory" (Paper presented at Psychology/Cognitive Neuroscience Colloquium, Hanover, N.H., April 27, 1990); Allman, *Apprentices of Wonder*, esp. pp. 72–75.

54. Marcus, "The Structure of Emotional Appraisal"; Annette Baier, "Getting in Touch with Our Feelings," *Topoi* 6 (1987): 89–97.

55. It is not appropriate here to raise the question of role of an ideological commitment to political egalitarianism in the public acceptance of Kantian ethics and other highly rationalistic theories of human thought. Cf. Ian Shapiro, *Political Criticism* (Berkeley: University of California Press, 1990).

56. Edelman, *Topobiology*.

57. Jacques Mehler, "Language Comprehension: The Influence of Age, Modality and Culture" (Paper presented at the 37th Annual Meeting of the Orton Dyslexia Society, Philadelphia; available from Audio-Stats Educational Services, Inglewood, Calif., tape no. 916R-21).

58. Siegfried Frey and Gary Bente, "Mikroanalyse Medienvermittelter Informationsprozesse: Zur Anwendung zeutreihen-basierter Notationsprinzipien auf die Untersuchung von Fernsehnachrichten," in *Massenkommunikation: Theorien, Methoden, Befunde*, ed. Max Kaase and Winfried Schulz (Opladen: Westdeutscher Verlag, 1989), pp. 508–26, esp. Figure 5; Gary Bente, Siegfried Frey, and Johannes Treeck, "Taktgeber der Informationsverarbeitung, *Medien Psychologie* 2 (1989): 137–60.

59. Lionel Tiger, "The Evolution of Cultural Norms," in *The Sense of Justice*, ed. Roger D. Masters and Margaret Gruter (Newbury Park, Calif.: Sage, 1992), pp. 278–89.

60. Cf. Robin Fox, *The Search for Society* (New Brunswick, N.J.: Rutgers University Press, 1989).

61. Edward T. Hall, *The Silent Language* (Greenwich, Conn.: Fawcett, 1959); Raymond L. Birdwhistell, *Kinesics and Context* (Philadelphia: University of Pennsylvania Press, 1970); Eibl-Eibesfeldt, *Human Ethology*.

62. L. J. Eaves, H. J. Eysenck, and N. G. Martin, *Genes, Culture, and Personality: An Empirical Approach* (New York: Academic Press, 1989).

63. C. Robert Cloninger, "A Unified Biosocial Theory of Personality and Its Role in the Development of Anxiety States," *Psychiatric Developments* 3 (1986): 167–226; and "A Systematic Method of Clinical Description and Classification of Personality Variants," *Archives of General Psychiatry* 44 (1987): 573–88. See also MacDonald, *Social and Personality Development*.

64. On the concept that specific areas in the central nervous system and the behaviors they control are "tuned" or modulated by a specific neurotransmitter, see David Servan-Schriber, Harry Printz, and Jonathan D. Cohen, "A Network Model of Catecholamine Effects: Gain, Signal-to-Noise Ratio, and Behavior," *Science* 249 (1990): 892–95; N. I. Syed, A. G. M. Bulloch, and K. Lukowiak, "In Vitro Reconstruction of the Respiratory Central Pattern Generator of the Mollusk *Lymnaea*," *Science* 250 (1990): 282–85.

65. Michael Stanley and J. John Mann, "Biological Factors Associated with Suicide," *Review of Psychiatry* 7 (1988): 334–52; Richard Wurtman and Judith Wurtman, "Carbohydrates and Depression," *Scientific American* 262 (1989): 68–75; Matti Virkkunen, Arno Nuutila, Frederick K. Goodwin, and Markku Linnoila, "Cerebrospinal Fluid Monoamine Metabolite Levels in Male Arsonists," *Archives of General Psychiatry* 44 (1987): 241–47. For a survey of the findings with regard to serotonin, see Roger D. Masters and Michael T. McGuire, eds., *The Neurotransmitter Revolution: Serotonin, Social Behavior, and the Law* (Carbondale: Southern Illinois University Press, 1993).

66. For dogs, see Matthew Margolis and Catherine Swan, *The Dog in Your Life* (New York: Vintage, 1979), pp. 47–48. On horses, see Hearne, *Adam's Task*. More detailed statistical analysis of the social behavior of vervet monkeys suggests that individual differences in the frequency of various social behaviors can be described in terms of three principal factors or dimensions (Michael T. McGuire, personal communication); these dimensions resemble those in Cloninger's model of personality.

67. Recent research points even more precisely to links between mutations, neurochemicals, and behavior. For associative learning of spatial tasks, a single gene mutation controlling a specific enzyme can produce animals with performance deficits (Alcino J. Silva, Charles F. Stevens, Susumu Tonegawa, and Yanyan Wang, "Deficient Hippocampal Long-term Potentiation in a-Calcium Calmodulin Kinase II Mutant Mice," *Science* 257 [1992]: 201–6; Alcino J. Silva, Richard Paylor, Jeanne M. Wehner, and Susumu Tonegawa, "Impaired Spatial Learning in a-Calcium Calmodulin Kinase II Mutant Mice," *Science* 257 [1992]: pp. 206–11).

68. Michael T. McGuire and Michael J. Raleigh, "Behavioral and Physiological Correlates of Ostracism," in *Ostracism: A Social and Biological Phenomenon*, ed. Margaret Gruter and Roger D. Masters (New York: Elsevier, 1986), pp. 39–52; Michael J. Raleigh and Michael T. McGuire, "Animal Analogues of Ostracism: Biological Mechanisms and Social Consequences," in *Ostracism: A Social and Biological Phenomenon*, ed. Margaret Gruter and Roger D. Masters (New York: Elsevier, 1986), pp. 53–66; Douglas Madsen, "A Biochemical Property Relating to Power-seeking in Humans," *American Political Science Review* 79 (1985): 448–57.

69. Robert M. Sapolsky, "The Endocrine Stress-Response and Social Status in the Wild Baboon," *Hormones and Behavior* 16 (1982): 279–92, "Individual Differences in Cortisol Secretory Patterns in the Wild Baboon: The Role of Negative Feedback Sensitivity," *Endocrinology* 113 (1983): 263–68, and "Stress in the Wild," *Scientific American* 262 (1990): 116–23. For similar evidence on human children, see Montagner, *L'Enfant et la communication*.

70. Robert Plomin and Denise Daniels, "Why Are Children in the Same Family So Different from One Another?" *Behavior and Brain Science* 10 (1987): 1–30; Robert Plomin, "The Role of Inheritance in Behavior," *Science* 248 (1990):183–88; Eaves, Eysenck, and Martin, *Genes, Culture, and Personality*.

71. See Lewontin, Rose, and Kamin, *Not in Our Genes*.

72. Benson Ginsburg, "Ontogeny, Social Experience, and Serotonergic Functioning," in Masters and McGuire, *The Neurotransmitter Revolution*, ch. 9.

73. Carroll E. Izard, Elizabeth A. Hembree, and Robin R. Huebner, "Infants' Emotion Expressions to Acute Pain: Developmental Change and Stability of Individual Differences," *Developmental Psychology* 213 (1987):105–13.

74. Jerome Kagan, J. Steven Reznick, Nancy Snidman, "Biological Bases of Childhood Shyness," *Science* 240 (1988):167–71.

75. Martha Denckla, "The Neurology of Social Competence," *ACLS Newsbriefs*, June/July 1986, p. 1.

76. Deane W. Lord, "Doctoral Studies with a Difference," *Harvard Alumni Gazette* 85 (1990):27, 30.

77. Glendon Schubert, "Introduction: Primatology, Feminism, and Political Behavior," in *Primate Politics*, ed. Schubert and Masters, chap. 1.

78. E.g., Carol Gilligan, *In a Different Voice* (Cambridge, Mass.: Harvard University Press, 1982); Robert Rosenthal and Donald B. Rubin, "Further Meta-

Analytic Procedures for Assessing Cognitive Gender Differences," *Journal of Educational Psychology* 74 (1982):708–12; Nicole A. Steckler and Robert Rosenthal, "Sex Differences in Nonverbal and Verbal Communication with Bosses, Peers, and Subordinates," *Journal of Applied Psychology* 70 (1985):157–63; Philip Shaver and Clyde Hendrick, eds., *Sex and Gender: Review of Personality and Social Psychology*, (Beverly Hills, Calif.: Sage, 1987), vol. 7; Ethel Klein, *Gender Politics* (Cambridge, Mass.: Harvard University Press, 1987).

79. John Money and Anke A. Ehrhardt, *Man and Woman, Boy and Girl: The Differentiation and Dimorphism of Gender Identity from Conception to Maturity* (Baltimore: Johns Hopkins University Press, 1972); Roberta L. Hall, ed., *Sexual Dimorphism in Homo Sapiens: A Question of Size* (New York: Praeger, 1982).

80. For a more detailed review of the gender differences described below, see Roger D. Masters, "Gender and Political Cognition," *Politics and the Life Sciences* 8 (1989):3–39.

81. Sarah Blaffer Hrdy, *The Woman Who Never Evolved* (Cambridge, Mass.: Harvard University Press, 1981); Frans de Waal, "Sex Differences in the Formation of Coalitions among Chimpanzees," in *Primate Politics*, ed. Glendon Schubert and Roger D. Masters (Carbondale: Southern Illinois University Press, 1991), chap. 6.

82. Compare Ruth Hubbard, Mary Sue Henefin, and Barbara Fried, eds., *Women Look at Biology Looking at Women* (Cambridge, Mass: Schenkman, 1979), with Lynda Birke, *Women, Feminism, and Biology: The Feminist Challenge* (Brighton, England: Harvester Press, 1986). See also Roberta L. Hall, *Male-Female Differences: A Bio-Cultural Perspective* (New York: Praeger, 1985).

83. Gilligan, *In a Different Voice*.

84. Judith A. Hall, "Gender Effects in Decoding Nonverbal Cues," *Psychological Bulletin* 85 (1978): 845–57, and "On Explaining Gender Differences: The Case of Nonverbal Communication," in Shaver and Hendrick, eds., *Sex and Gender*, pp. 177–200.

85. Compare Wayne A. Babchuck, Raymond B. Hames, and Ross A. Thomason, "Sex Differences in the Recognition of Infant Facial Expressions of Emotion: The Primary Caretaker Hypothesis," *Ethology and Sociobiology* 6 (1985):89–102, with Masters, "Gender and Political Cognition."

86. Patricia R. Barchas, William A. Harris, William S. Jose II, and Eugene A. Rosa, "Social Interaction and the Brain's Lateralization of Hemispheric Function," in *Social Cohension: Essays toward a Sociophysiological Perspective* ed. Patricia R. Barchas and Sally P. Mendoza (Westport, Conn.: Greenwood, 1986), pp. 131–50; Masland, "Neurological Aspects of Dyslexia."

87. Doreen Kimura and Elizabeth Hampson, "Neural and Hormonal Mechanisms Mediating Sex Differences in Cognition," *Research Bulletin*, no. 689, Department of Psychology, University of Western Ontario, London, Ontario, Canada; Doreen Kimura, "Sex Differences in the Brain," *Scientific American* 267 (September 1992):119–25.

88. Masters, "Gender and Political Cognition"; Roger D. Masters and Stephen J. Carlotti, Jr., "Gender Differences in Responses to Political Leaders," in *Social Stratification and Socioeconomic Inequality* ed. Lee Ellis (Boulder, Colo.: Praeger, in press).

89. In more technical language, experiments need to be designed using stimuli that activate established cognitions and emotions with degrees of arousal comparable to those in socially significant situations. Insofar as associative learning

involves forming or strengthening a network of independent brain modules in response to a specific cue, artificial situations or stimuli may not reveal the full complexity of the way individuals feel and learn. Each individual's own kin or personal friends, while generating deeply felt feelings and attitudes, would vary from one experimental subject to another, whereas relationships created in the laboratory cannot elicit long-standing memories of events, attitudes, and behavior. The only alternative might be figures known from the world of entertainment and sport—but here differences in taste (including those of typical male and female interests) would confound any study of the way any single individual elicits emotions and thoughts across the entire population.

90. Hans Mathias Kepplinger, "The Impact of Presentation Techniques: Theoretical Aspects and Empirical Findings," in *The Psychological and Semiotic Processing of Televised Political Advertising* ed. Franc Biocca (Hillsdale, N.J.: Lawrence Erlbaum, 1990).

91. Roger D. Masters, Siegfried Frey, and Gary Bente, "Dominance and Attention: Images of Leaders in German, French, and American TV News," *Polity* 25 (Spring 1991): 373–94.

92. For reviews of the research program, see Lanzetta, Sullivan, Masters, and McHugo, "Viewers' Emotional and Cognitive Responses to Televised Images of Political Leaders"; Roger D. Masters and Denis G. Sullivan, "Nonverbal Behavior and Leadership: Emotion and Cognition in Political Information Processing," Working Paper no. 90-4, Institute of Governmental Studies (Berkeley: University of California, 1990). In our experiments, subjects were merely told that we were interested in the effects of the media on politics (and generally did not suspect our interest was in nonverbal behavior). After answering a standard questionnaire, indicating political opinions and attitudes toward leaders as well as other background information, subjects were presented with videotaped excerpts (20 to 120 seconds in length) showing close-up images of leaders. After each excerpt, viewers typically described the nonverbal behavior of the leader along a number of dimensions (the leader was strong, happy, angry, fearful, etc.), using 0–6 scales to provide measure of perceived display intensity. Then viewers were asked to report on their own emotional feelings during the excerpt, again using 0–6 scales for different emotions (happy, angry, afraid, etc.). Since emotions that are described verbally (e.g., happiness, anger) correspond to specific patterns of physiological response, verbal self-reports of how the subject felt during an excerpt were validated by comparing them with known psychophysiological measures of emotional arousal (McHugo, Lanzetta, Sullivan, Masters, and Englis, "Emotional Reactions to Expressive Displays of a political leader"). Finally, at the end of most of our studies we measured changes in attitude to see whether the excerpts had changed the viewers' opinions of the leaders. In some cases, a study used a single leader (President Reagan) exhibiting the three types of display, with systematic variations in the channel of communication (sound plus image, image only, sound only, filtered sound plus image, or text only); in addition to studies using paper-and-pencil responses in three or more of these media conditions, experimental evidence of psychophysiological effects was used to verify the accuracy of verbal self-reports of emotion.

In another set of experiments, all candidates during the presidential campaigns of 1984 and 1988 were shown in two excerpts, one neutral and the other the best happy/reassuring display available early in the campaign year; identical displays were shown to groups of subjects before the first primary election and at the end of the campaign, just before election day (Sullivan and Masters, "Happy Warriors"; Denis G. Sullivan and Roger D. Masters, "Nonverbal Behavior, Emotions, and Democratic Leadership," in *Reconsidering Democracy* ed. George Marcus and John Sullivan [in press]; Denis G. Sullivan and Roger D. Masters, "Viewers' Emotional and Cognitive Responses to Candidates' Expressive Displays in the 1988 Presidential Election" [Paper presented to annual meeting of International Society for Political Psychology, Washington, D.C., July 12, 1990]). Another study inserted silent excerpts of Reagan's display behavior in the background of routine TV news stories so that different groups of subjects saw identical news accompanied by different mixtures of silent facial display cues (Sullivan and Masters, "Biopolitics, the Media and Leadership: Nonverbal Cues, Emotions and Trait Attributions in the Evaluation of Leaders"). Finally, the American studies using the three types of display were replicated in France, using excerpts of three national leaders just before the legislative elections of 1986 (Roger D. Masters and Jean Mouchon, "Les gestes et la vie politique"; Masters and Sullivan, "Nonverbal Displays"; Roger D. Masters and Denis G. Sullivan, "Facial Displays and Political Leadership in France," *Behavioral Processes* 19 (1989):1–30.

93. Although these feelings were originally recorded using eight different verbal scales, these self-reported scores reflect the two underlying dimensions of emotion that have been consistently observed in other studies: one usually called "positive" or "hedonic" (reassuring, affiliative, and comforting) and the other "negative" or "agonic" (competitive, aggressive, fearful) (Michael R. A. Chance, "Attention Structures as the Basis of Primate Rank Orders," in *The Social Structure of Attention*, ed. Michael R. A. Chance and Ray R. Larson [New York: John Wiley, 1976]; Michael R. A. Chance, ed., *Social Fabrics of the Mind* ([Hillsdale, N.J.: Lawrence Erlbaum, 1989]). The original eight scales of emotional response (inspired, joyful, comforted, interested, angry, disgusted, fearful, confused) were largely redundant (see Masters et al., "Facial Displays of Leaders," esp. Tables 7 and 8); as a result, more recent studies have used only four scales (happy, comforted, angry, afraid). The pattern of emotions observed in our studies was parallel to that found to predict voting behavior in national public opinion polls, was stable regardless of the media condition used in the experiment, and had the same structure in both France and the United States (cf. Robert P. Abelson, Donald R. Kinder, M. D. Peters, and Susan T. Fiske, "Affective and Semantic Components in Political Person Perception," *Journal of Personality and Social Psychology* 42 [1982]:619–30; George Marcus, "The Structure of Emotional Appraisal").

94. In addition to references cited above, see James S. Newton, Roger D. Masters, Gregory J. McHugo, and Denis G. Sullivan, "Making Up Our Minds: Effects of Network Coverage on Viewer Impressions of Leaders," *Polity* 20 (1987):226–46. As an illustration of these complexities, the table below presents evidence from a study that shows differences between responses of blacks and whites. Throughout our studies, emotional responses to most leaders' nonverbal displays have translated to attitude change. In a study just before the 1988 election, however, emotions felt while watching presidential candidates' display behavior were less likely to influence the emotions of black viewers than those of whites. Although the blacks had perceived the difference between neutral and happy displays and had responded with very strong positive emotions when they saw Jackson and Dukakis, their responses to all leaders—including those they preferred—were highly cognitive and uninfluenced by emotion (Roger D. Masters, "Ethnic and Personality Differences

in Response to TV Images of Leaders" [Paper presented at the 1991 meeting of the American Political Science Association, Washington, D.C., August 29, 1991]).

Pretest Attitudes and Emotional Responses to Nonverbal Behavior as Factors Influencing Post-Attitude

	N	Pretest Attitude	Net Warmth to		Party Identification	Ideology	Assessed Leadership	Issue Agreement	Adjusted R^2
			H/R Display	Neutral Display					
November, 1988									
Bush									
Blacks	47	.56	.89	.71	−1.00	.30	1.36	2.44	.63
Whites	151	.55	.65	.32	.25	−.64	.80	2.45	.78
Dukakis									
Blacks	48	.55	−.02	.50	−2.93	−3.24	−0.18	1.93	.47
Whites	152	.42	1.28	.12	−.47	.69	.93	2.48	.74
Dole									
Blacks	49	.43	.91	1.26*	.42	2.19	4.70	−1.36	.46
Whites	151	.53	1.28	.39	1.51	.33	.10	1.20	.42
Jackson									
Blacks	50	.52	.54	.16	.87	−1.11	1.53*	4.51	.58
Whites	151	.35	.30	1.58	−1.49*	−.11	2.15	2.04	.76
Robertson[a]									
Whites	147	.41	.85	.50*	.71	.48	1.31	1.21	.45
Reagan[a]									
Whites	152	.63	−.06	.90	.98	.95	1.56	1.62*	.81
Hart									
Blacks	49	.25*	−.11	.70	−3.06*	3.53	5.70	1.61	.50
Whites	150	.35	1.10	1.41	.78	3.16	.80	4.09	.51
Studies in 1984 with Identical Excerpts (Predominantly White Samples)									
Reagan									
Jan.	75	.52	1.22	.90	−3.45	—	3.91	−2.00	.84
Oct.	90	.22	2.01	.41	−.75	—	3.56	.41	.76
Hart									
Jan.	75	.45	1.21	1.22	.68	—	.90	1.11	.43
Oct.	88	.61	1.43	−.45	−.05	—	2.96	.19	.69

Note: Cell entries are unstandardized multiple regression coefficients.
$p < .01$ = underlined and bold; $p < .05$ = bold; $p < .10$ = asterisk.
Source for 1984 data: Sullivan and Masters, "Happy Warriors."
[a]Equations for black responses to Reagan, Gephardt, and Robertson, and white responses to Gephardt were not computed because in these cases there was no significant change between pretest and posttest attitudes (see n. 11).

6. Is Science Relative?

1. E.g., Larry Laudan, *Science and Relativism: Some Key Controversies in the Philosophy of Science* (Chicago: University of Chicago Press, 1990), pp. 159–60 et passim.
2. David Bloor, *Knowledge and Social Imagery* (London, Routledge, 1976), p. 3.
3. This argument is reinforced by the history of the conceptual innovations

underlying modern science that was sketched in the notes to chap. 2. The sequence of pagan antiquity, followed by Christianity and the gradual secularization of the later Middle Ages and Renaissance, was as unique as the constellation of socio-economic, religious, and cultural beliefs surrounding the emergence of modern scientific theories and practices.

4. See also Karl Mannheim, *Ideology and Utopia* (New York: Harcourt Brace, 1966). This point could easily be confirmed for intuition as well as for verification and pattern matching by pointing to the role of the shaman or priest in hunter-gatherer societies, which presumably resemble the earliest recognizably human cultures.

5. For R. A. Fisher's statistical principles, the classic statement is "The Mathematics of the Lady Tasting Tea," in Fisher, *The Design of Experiments* (Adelaide: Hafner Press, Macmillan, 1971). His role in the eugenics movement is described in Daniel Kevles, *In the Name of Eugenics* (New York: Knopf, 1985).

6. E. g., Isaac Asimov, *A Short History of Biology* (Garden City, N.Y.: Natural History Press, 1964); Peter J. Bowler, *Evolution: The History of an Idea* (Berkeley: University of California Press, 1984).

7. Werner Heisenberg, *Physics and Philosophy* (New York: Harper Torchbooks, 1958); Fritjof Capra, *The Tao of Physics* (Berkeley: Shambala, 1975). See below, chap. 7.

8. Stephen J. Gould, *Ontogeny and Phylogeny* (Cambridge, Mass.: Harvard University Press, 1977), p. 16.

9. Stephen Jay Gould and Niles Eldredge, "Punctuated Equilibria: The Tempo and Mode of Evolution Reconsidered." See also Stephen Jay Gould, "Punctuated Equilibrium in Fact and Theory," in *The Dynamics of Evolution*, ed. Albert Somit and Steven A. Peterson (Ithaca, N.Y.: Cornell University Press, 1992), pp. 54–84; and Niles Eldredge, "Punctuated Equilibria, Rates of Change, and Large-Scale Entities in Evolutionary Systems" (ibid., pp. 103–21).

10. The following discussion is based on Roger D. Masters, "Gradualism and Discontinuous Change in Evolutionary Theory and Political Philosophy" in *The Dynamics of Evolution*, ed. Albert Somit and Steven A. Peterson (Ithaca, N.Y.: Cornell University Press, 1992), pp. 282–319.

11. "Empedocles accounts for the creation of animals in the time of his Reign of Love, saying that 'many heads sprang up without necks,' and later on these isolated parts combined into animals" (Aristotle, *Generation of Animals*, 1.18.722b; in *Collected Works of Aristotle*, vol. 1, p. 1122. Cf. Gould, *Ontogeny and Phylogeny*, p. 413. Although Empedocles' own works have not survived, Aristotle's frequent citations of his work provide the best firsthand evidence of his theories. As will be noted below, confirmation of this account can be gained from Lucretius and the Epicurean tradition. Since Aristotle cites Empedocles in order to criticize his biological views, whereas Lucretius calls Empedocles a genius and endorses his theory, the combined references are likely to be reliable. Mere accident gave rise to different living forms, most of which—like a "man-faced ox-progeny"—perished; those animals that were adapted to the world survived, as a result of what Aristotle calls "necessity." Before criticizing Empedocles' biology, Aristotle presents the argument for it as follows:

> Why should not nature work, not for the sake of something, nor because it is better so, but just as the sky rains, not in order to make the corn grow, but of necessity? (What is drawn up must cool, and what has been cooled must

become water and descend, the result of this being that the corn grows.) Similarly if a man's crop is spoiled on the threshing-floor, the rain did not fall for the sake of this—in order that the crop might be spoiled—but that result just followed. Why should it not be the same with the parts in nature, e.g. that our teeth should come up of necessity—the front teeth sharp, fitted for tearing, the molars broad and useful for grinding down the food—since they did not arise for this end, but it was merely a coincidence result; and so with all other parts in which we suppose that there is purpose? Wherever then all the parts came about just as they would have been if they had come to be for an end, such things survived, being organized spontaneously in a fitting way; whereas those which grew otherwise perished and continue to perish, as Empedocles says his "man-faced oxprogeny" did. (*Physics*, 2.8.198b; *Collected Works of Aristotle*, vol. 1, p. 339)

Darwin cites this passage explicitly in the "historical sketch" that he added at the outset of *Origin of Species* (New York: Modern Library, n.d.), xiii*n*.) As is evident from Darwin's comments on the passage, he did not understand the context and misinterpreted Aristotle's point. In good part this is because, in this passage as elsewhere, Aristotle distinguishes the concepts of "necessity," "chance," and "nature" in a way that differs fundamentally from modern usage: what he calls chance is an unintended or accidental result within the domain of purposive behavior. Since Aristotle treats nature as a purposive realm, many phenomena that are today called "natural" or "chance" were, for Aristotle, merely "necessary." See Roger D. Masters, "Evolutionary Biology and Natural Right," in *The Crisis of Liberal Democracy*, ed. Kenneth Deutsch and Walter Soffer (Albany, N.Y.: SUNY Press, 1987). Many modern commentators have misunderstood Aristotle profoundly because they have failed to attend to this crucial difference between his terms and those of our own day.

12. While crude by comparison with contemporary biological theory, in part because of the simplicity of his image of "hopeful monsters," Empedocles provides us with an early instance of nongradual change as the basis of speciation. The images of long extinct forms of animals in Gould's *Wonderful Life* (New York: Norton, 1989) suggest, moreover, that the apparently bizarre character of nonsurviving forms may have been properly suggested by Empedocles's description of a "man-faced ox."

13. Victor Goldschmidt, *Le doctrine d'Epicure et le droit* (Paris: J. Vrin, 1977).

14. Lucretius, *On the Nature of Things*, ed. R. E. Latham (Harmondsworth, England: Penguin, 1952). All translations in the text are from this edition. Lucretius explicitly praises Empedocles as a "divine genius" (1.716–733). For Lucretius, the world as we see it is relatively new, yet fundamental change in the cosmos is imminent; like Empedocles, Lucretius does not attribute the beings we see to a beneficent divine plan (5.324–50; p. 181). During an early phase of the earth's history, natural causes produced a variety of living forms, but only those that successfully reproduced have survived (5.837–56; p. 197). "In those days the earth attempted also to produce a host of monsters, grotesque in build and aspect—hermaphrodites, halfway between the sexes yet cut off from either, creatures bereft of feet or dispossessed of hands, dumb, mouthless brutes, or eyeless and blind, or disabled by the adhesion of their limbs to the trunk, so that they could neither do anything nor go anywhere nor keep out of harm's way nor take what they needed. These and other such monstrous and misshapen births were created. But all in vain. Nature

debarred them from increase . . . In those days, again, many species must have died out altogether and failed to reproduce their kind" (*De Rerum Natura*, 5.837–56; p. 197). Although there was a historical period of wide variation in living forms, Lucretius explicitly denies that there are interspecific chimera in his own epoch on the grounds that "each species develops according to its own kind, and they all guard their specific characters" (5.922–24; p. 199). After each species is formed, their conservation is guaranteed by the divergent "nature" of each kind of animal: Lucretius is, like Empedocles, a theorist whose concept of time resembles recent models of punctuated equilibria.

15. As Lucretius concludes, "not only such arts as sea-faring and agriculture, city walls and laws, weapons, roads and clothing, but also without exception the amenities and refinements of life, songs, pictures, and statues, artfully carved and polished, all were taught gradually by usage and the active mind's experience as men groped their way forward step by step. So each particular development is brought gradually to the fore by the advance of time, and reason lifts it into the light of day" (*De Rerum Natura*, 1448–59; p. 216). A similar gradualist view of human history, according to which all social rules arose through consent, is also presented in the Epicurean "genealogy" of ethics by Hermarque. See *Letters on Empedocles*, cited in Goldschmidt, *Le doctrine d'Epicure et le droit*, pp. 287–97, esp. p. 293–94.

16. Like Aristotle, Hobbes usually takes contemporary animal species as given; when he discusses the "natural condition of mankind," Hobbes refers to a currently existent state of affairs rather than the historical origin of the human condition ("it was never generally so, over all the earth" [*Leviathan*, 1.13, p. 101]). Whereas ancients like Lucretius consider the discovery of language, the arts, and political institutions as a slow, incremental process, Hobbes treats each step as if it were a sudden event. Language, for example, is an "invention" whose origin can be compared to the invention of alphabetic writing and printing (*Leviathan*, 1.4, p. 33). The founding of the state is a "covenant" that is described as a specific event (*Leviathan*, 2.17–18, pp. 129–41), be it violent conquest (a "commonwealth by acquisition") or mutual consent (a "commonwealth by institution"). His theory is strangely constructed so that its teaching and acceptance are formally necessary to its success, producing lasting control over the tendency of all previous societies to collapse into civil war (*Leviathan*, 2.20, pp. 157–58; 30, pp. 248–49; 31, p. 270). The key to such a radical change in the human condition lies in the paradoxical character of Hobbes's teaching, since the universal adoption of his way of thinking is necessary and sufficient as a means of totally transforming political life. See Roger D. Masters, "Hobbes and Locke," in Ross Fitzgerald, ed., *Comparing Political Thinkers* (Sydney and New York: Pergamon Press, 1980), pp. 116–40. In technical terms, Hobbes's theory of the rights of nature and the laws of nature entails "reflexive predictions" whose outcome depends on their "dissemination status"; see Roger Buck, "Reflexive Predictions," *Philosophy of Science* 30 (1963):359–69.

17. Marx explicitly endorsed the broad principles of organic evolution presented by Darwin, but the relationship between his theory and contemporary biology has been controversial. For some, Marx would have favored the sociobiology of Edward O. Wilson and Richard Alexander; for others he would have roundly condemned it. Cf. Lewis Feuer, "Marx and Engels as Sociobiologists," *Survey* 23 (1977), pp. 109–36; W. J. M. Mackenzie, *Biological Ideas in Politics* (New York: St. Martin's Press, 1979), pp. 50–52; Robert Tucker, *Philosophy and Myth in Karl*

Marx (Cambridge: University Press, 1961). The complexity arises because, although Marx set himself the thoroughly modern goal of unifying the natural and social sciences (Marx, "Private Property and Communism," in *Economic and Philosophic Manuscripts of 1844*, ed. Dirk J. Struik, [New York: International Publishers, 1964], p. 143), he was steeped in the philosophy and science of classical antiquity. As a result of writing his doctoral dissertation on the Greek atomists Democritus and Epicurus, the young Marx was particularly impressed by Lucretius as the model of "a mind free and sharp," willing to describe "Nature without gods, gods without a world" (Mackenzie, *Biological Ideas*, p. 54).

Although Marx did not entirely accept the view of biological and historical change presented in Lucretius' *On the Nature of Things*, this remark reflects Marx's primary objective: to replace the theological account of human existence with a more naturalistic one capable of unifying the natural and social sciences. Cf. the note to Marx's doctoral dissertation entitled "Reason and the Proofs of God," in *Writings of the Young Marx on Philosophy and Society*, ed. Loyd D. Easton and Kurt H. Guddat (Garden City, N.Y.: Anchor, 1967), pp. 64–66. Marx's insistence that human beings have evolved or "developed" by a process of "natural history" probably explains his enthusiasm for Darwinian theory, since the specific mechanisms of evolutionary change described by Darwin struck Marx as a vulgar illustration of British bourgeois thought; for Marx, Darwin's view of natural selection was little more than a Hobbesian "war of all against all," conveniently transposing to nature the ideological norm and practical reality of competitive capitalism.

Even granted these defects, however, Darwin was invaluable for Marx because his *Origin of Species* could serve, as did Lucretius's *On the Nature of Things*, as an ally against religious dogma. Thus, when Engels says that Marx's "fundamental proposition . . . is destined to do for history what Darwin's theory has done for biology" (1888 Preface, *Communist Manifesto*, ed. Dirk J. Struik [New York: International Publishers, 1971], pp. 136–37), the reference is not to Darwin's gradualist model of rates of change. Marx saw himself as engaged in a battle to demonstrate that God could no longer be used to explain either nature or human society ("Private Property and Communism," *1844 Manuscripts*, pp. 145–46). By 1844, Marx's concern had already focused on the political goals of a fundamental transformation of the human condition, and it was from this perspective that he sought a materialist, scientifically objective account of human history.

18. Using a Hegelian term, Marx speaks of humans as "a species being"—i.e., a being capable of "free" or "conscious life activity" ("Estranged Labor," *1844 Manuscripts*, p. 113)—and postulates a chasm separating humans from other animals.

> In creating a world of objects by his practical activity, in his work upon inorganic nature, man proves himself a conscious species being . . . Admittedly animals also produce. They build themselves nests, dwellings, like the bees, beavers, ants, etc. But an animal only produces what it immediately needs for itself or its young. It produces one-sidedly, whilst man produces universally. It produces only under the dominion of immediately physical need, whilst man produces even when he is free from physical need and only truly produces in freedom therefrom. An animal produces only itself, whilst man reproduces the whole of nature. An animal's product belongs immediately to its physical body, whilst man freely confronts his product. An animal forms things in accordance with the standard and the need of the species to which it belongs, whilst man

knows how to produce in accordance with the standard of every species, and knows how to apply everywhere the inherent standard to the object ("Estranged Labor," *1844 Manuscripts*, pp. 113–14)

This is not an isolated passage in the writings of the young Marx. In the *German Ideology*, written with Engels in 1845–46, we find the same radical distinction between humans and other species (*German Ideology*, ed. C. J. Arthur [New York: International Publishers, 1970], pt. 1, p. 51). Given the uniqueness of human beings, Marx pays relatively little attention to the precise pattern of human evolution beyond condemning the doctrine of divine creation. The attempt to account for the origin of species is ultimately circular, since it is impossible to go beyond the Aristotelian account of parent–offspring resemblance without becoming lost in the "abstraction" of thinking of "man and nature as non-existent." Since Marx sees modern geology as replacing the biblical teaching of the "creation of the earth" by a "process" that is equivalent to "self-generation," the emergence of species might just as well be due to spontaneous generation: "*Generatio aequivoca* is the only practical refutation of the theory of creation" ("Private Property and Communism," *1844 Manuscripts*, p. 144).

19. "The social reality of nature, and human natural science, or the natural science about man, are identical terms" ("Private Property and Communism," *1844 Manuscripts*, p. 145). Humans can and soon will achieve a radically humanistic appropriation of all "nature." The young Marx seems to have adopted a secular version of the Christian doctrine of "special creation," albeit one that views the human species as its own creator. Cf. Tucker, *Philosophy and Myth in Karl Marx*.

20. In both the *German Ideology* and the *Communist Manifesto*, Marx and Engels give a detailed account of the "revolutionary" transformation of feudal society into capitalist epoch in which the bourgeoisie bases its wealth and power on control over the industrial mode of production. Apart from a few asides on the likely future of England, Marx generally treats the transformation of capitalist society into communism as a process that will require a political revolution ("Bourgeois and Proletarians," *Communist Manifesto*, p. 101). The final transition to communism requires a particularly radical step because it must be worldwide in scope "as the act of the dominant peoples 'all at once' and simultaneously" (*German Ideology*, pt. 1, p. 56). This global upheaval is possible because, in the process of establishing a single world market, capitalism destroys all class distinctions except the contradiction between the bourgeoisie and the proletariat. In the *1844 Manuscripts* as well as the *Critique of the Gotha Programme* (1875), moreover, Marx intimates that the transition from capitalism to communism will pass through a brutal stage of "crude communism" before ushering in the definitive freedom of mankind ("Private Property and Communism," *1844 Manuscripts*, p. 133–35). See also the Preface to the *Contribution to the Critique of Political Economy*, ed. L. Colletti (New York: Vintage, 1975), pp. 425–26; *German Ideology*, pt. 1, pp. 59, 87–89, 94–95).

21. Whereas all prior history can be described as "natural," since changes occurred in an involuntary manner, the next phase of history will permit "the control and conscious mastery of these powers, which, born of the action of men on one another, have till now overawed and governed men as powers completely alien to them (*German Ideology*, pt. 1, p. 55). By abolishing the division of labor and private property, communism thus frees humans not only from exploitation by others, but

from the domination of natural necessity. Cf. "Private Property and Communism," *1844 Manuscripts*, p. 135–44.

22. Aristotle's biological writing is focused on the functional adaptation of traits rather than their evolutionary origin and change; although he is more interested in reproduction and embryos than in the origin of species, it does not follow—as is widely believed—that he denies the possibility of evolutionary change. As one careful student of Aristotle's concept of *eidos* (form or type) concludes: "I believe that Aristotle is not committed to Noah's Ark Essentialism, to 'typology,' or to put it most paradoxically, he is not committed to the taxonomic theory which is sometimes called Aristotelian Essentialism" (Anthony Preus, "*Eidos* as Norm in Aristotle's Biology," *Nature and System* 1 (1979):81. It may well be asked why has there been such a massive misunderstanding of Aristotle. First, Aristotle's goal in such works as *Parts of Animals*, *History of Animals*, and *Generation of Animals* is to establish biology as an empirical science. In particular, he seeks to dismiss the approach to classifying species, found, for example, in Plato's *Statesman*, that is based on simple "dichotomies" (*Parts of Animals*, 1.2–4.642b–644b). In so doing, Aristotle's usual method can be described, in contemporary terms, as "phenetic" classification—i.e., he bases his concept of species on visible or phenotypical traits rather than on genetic origin.

Second, and equally important, Aristotle is not particularly interested in the "origin of species." Because his primary concern is the relationship between form and purpose, Aristotle seeks to explain what we now call adaptation or function as a natural process, making living beings amenable to scientific analysis. While often speaking of species as if they had existed from all time, Aristotle is quite open to the emergence of new forms (e.g., *Generation of Animals*, 2.7 746b) as well as to the existence of intermediate forms inconsistent with what Preus calls "Noah's Ark Essentialism" (*Parts of Animals*, 4.5 681a), cited by Preus, "*Eidos*," p. 81). When Aristotle speaks of the *telos* or end of a structure, he is not presuming that animals demonstrate a divine plan of the sort accepted in the Judeo-Christian tradition; rather, Aristotle seeks empirical reasons for the observed shapes and characteristics of living beings on the assumption that only such "final causes" would explain the difference between animate and inanimate beings (e.g., *Physics*, 2.8.198b–199a). Aristotle's concept of functionalism is thus close to what is today called "teleonomy." See Colin Pittendrigh, "Adaptation, Natural Selection, and Behavior," in *Behavior and Evolution*, ed. Anne Roe and George Gaylord Simpson (New Haven, Conn.: Yale University Press, 1958), pp. 390–416; Ernst Mayr, "Teleological and Teleonomic: A New Analysis," *Boston Studies in the Philosophy of Science* 14 (1974):91–117.

Hence, when Aristotle refers to individual development as a recapitulation of phylogeny, it is to "reinforce his belief in the epigenetic nature of development" and not part of an elaborate theory of evolution (Gould, *Ontogeny and Phylogeny*, p. 16). As a result, it is a serious error to confuse Aristotelian biology with later theological uses of teleology. In accounting for the generation of animals, Aristotle notes "that it is thought that all animals are generated out of semen, and that the semen comes from the parents. That is why it is part of the same inquiry to ask whether both male and female produce it or only one of them, and to ask whether it comes from the whole of the body or not from the whole" (*Generation of Animals*, 1.17.721b; *Collected Works of Aristotle*, vol. 1, p. 1120).

There were sound intellectual reasons for Aristotle's focus, since he confronted objections to a scientific biology that were the opposite of those facing Darwin. In the mid-nineteenth century, the main obstacle to a scientific study of living beings came from the theological doctrine that all animals were the result of God's plan of Creation. For Aristotle, the issue arose because earlier thinkers had attributed everything to chance and thus denied that what we call a species is "natural." More specifically, Aristotle was concerned to show that the theories of Empedocles could not explain the transmission of traits from parents to offspring and hence were unsatisfactory as a foundation for biology. Confronted by such uncertainty, Aristotle was forced to assess questions like the inheritance of acquired characteristics (e.g., "the argument that mutilated parents produce mutilated offspring"), which would be evidence that the entire soma or body of both parents is the source of reproductive material.

23. "First, then, the resemblance of children to parents is no proof that the semen comes from the whole body, because the resemblance is found also in voice, nails, hair, and way of moving, from which nothing comes. And men generate before they have certain characters, such as a beard or grey hair. Further, children are like their more remote ancestors from whom nothing has come, for the resemblances recur at an interval of many generations, as in the case of the woman in Elis who had intercourse with a negro; her daughter was not negroid but the son of that daughter was" (*Generation of Animals*, 1.18.722a; *Collected Works*, vol. 1, p. 1121).

For the generation is for the sake of the substance, and not this for the sake of the generation. Empedocles, then, was in error when he said that many of the characters presented by animals were merely the results of incidental occurrences during their development; for instance, that the backbone is as it is because it happened to be broken owing to the turning of the foetus in the womb. In so saying he overlooked the fact that propagation implies a creative seed endowed with certain powers. Secondly, he neglected another fact, namely, that the parent animal pre-exists, not only in account, but actually in time. For man is generated from man; and thus it is because the parent is such and such that generation of the child is thus and so. (*Parts of Animals*, 1.1.640a; *Collected Works*, vol. 1, p. 996)

Cf. *Politics*, 2.3.1262a. Aristotle's emphasis on these aspects of reproduction has thus led to the suggestion that he be credited with the discovery of the principle of DNA (Max Delbruck, "Aristotle-totle-totle," in *Of Microbes and Life*, ed. J. Monod and E. Borek, [New York: Columbia University Press, 1971], pp. 50–55, and "How Aristotle Discovered DNA," in *Physics and Our World: A Symposium in Honor of Victor F. Weisskopf* ed. Kerson Huang, [New York: American Institute of Physics, 1977], pp. 123–30).

24. "Now that *this* is impossible is plain, for neither would the separate parts be able to survive without having any soul or life in them, nor if they were living things, so to say, could several of them combine so as to become one animal again. Yet those who say that semen comes from the whole of the body really have to talk in that way, and as it happened during [Empedocles's] 'Reign of Love', so it happens according to them in the body" (*Generation of Animals*, 1.18.722b). Observable phenomena in ontogeny, rather than speculative phylogeny, provide Aristotle with basic insights into the nature of animals. A similar contrast between Aristotle and Empedocles is equally evident in their accounts of gender (*Generation*

of Animals, 1.18.722b). Although Aristotle's understanding of the role of males and females in procreation departs from our own, he sought empirical evidence from ontogeny to explain how offspring derive physical traits from their parents. The difference between these ancient thinkers and contemporary biology lies less in the basic concepts of ontogeny and phylogeny than in their theoretical articulation. For Empedocles, a phylogenetic account structurally similar to neo-Darwinian evolutionary theory (random or accidental variation culled by natural selection) was combined with a view of ontogenesis based on somatic inheritance and the transmission of acquired traits. For Aristotle, stress on the functional unity and adaptation of organisms led to concern for mechanisms of inheritance, while phylogeny was either ignored or dismissed in favor of the view that all species form a natural continuum of complexity; what has mistakenly been described as "typological" thinking in Aristotle is actually his insistence on treating the organism as an adapted, functioning offspring of a naturally distinct and identifiable species.

Modern biology has, of course, combined phylogenetic views like those of Empedocles with ontogenetic principles akin to those of Aristotle. The difference between Empedocles and Aristotle thus points to two quite different ways of understanding species. For Empedocles, as later for Darwin, the accidental causes of speciation mean that the classification of species is a human artifact. In contrast, Aristotle would have agreed with a modern systematist like Mayr, for whom species exist in nature even if the boundaries of the species concept cannot always be defined with mathematical rigor. Cf. Mayr, *Animal Species and Evolution*, p. 29.

25. Just as each animal species has a "way of life" that influences individual and social behavior, so different human populations have a characteristic "way of life" with social and political consequences (*Politics*, 1.1–10 [esp. 8], 1256a–b). Moreover, each political community can be described as having a regime or constitution of a determinate species—the famous classification of six types of regime, three "good" and three in the interest of the ruling class (*Politics*, 3.7–9; 4.1–14). The same concept that Aristotle uses to refer to kinds of animals (*eidos*) is used to refer to the kinds of political order found in various human societies. See Roger D. Masters, "Nature, Human Nature, and Political Thought" in *Human Nature in Politics*, ed. Roland Pennock and John Chapman (New York: New York University Press, 1977); Preus, "*Eidos*"; and Larry Arnhart, "Aristotle, Chimpanzees, and Other Political Animals," *Social Science Information* 19 (1990):479–559.

26. When describing the origin of civilized societies like the city-state (*polis*) of the Greeks, Aristotle summarizes briefly an account that can only be characterized as "evolutionary" (*Politics*, 1.1–2.1252a–1253a); at no time does Aristotle imply that there has been—or will be—a sudden and irreversible quantum change in the human condition. When explicitly analyzing political upheavals, he describes "revolutions" as changes in regime that can be more or less extensive, but in no case are they presented as irreversible (*Politics*, 5.1–12, esp. 1.1301b). As a result, the best practicable regime for most instances, a "polity," is a mixed form of government that moderates the conflicts that characterize most communities. For Aristotle's political teaching, see Leo Strauss, *The City and Man* (Chicago: Rand McNally, 1964); Steven Salkever, *Aristotle on the Mean* (Princeton, N.J.: Princeton University Press, 1990); Arlene W. Saxonhouse, *Fear of Diversity: The Birth of Political Science in Ancient Greek Thought* (Chicago: University of Chicago Press, 1992), chaps. 8, 9.

27. Something like natural selection operates on all wild animals; human evolu-

tion required "multitudes of centuries" (*Second Discourse*, in *Collected Writings of Rousseau*, ed. Roger D. Masters and Christopher Kelly [Hanover, N.H.: University Press of New England, 1992], vol. 3, pt. 2, p. 45) and is a process in which "the lapse of time compensates for the slight probability of events" due to the "surprising power of very trivial causes when they act without interruption" (*Second Discourse*, pt. 1, pp. 42–43). Elsewhere Rousseau indicated that the historical changes producing society were "determined by the climate and by the nature of the soil," factors that can be traced to the inclination of the earth's axis and the resulting seasonal variations in climate (*Essay on the Origin of Languages*, chap. 9, pp. 38–39). To be sure, Rousseau saw unique principles underlying human history. Whereas other species behave on the basis of natural "instincts," humans are "free"—at least in the sense that they can "perfect" their own behavior and thereby change the way that they adapt to the world (*Second Discourse*, pt. 1, pp. 26–27). But this difference does not seem to rest on special creation or other extraevolutionary phenomena. Because he accepted Buffon's observation that the life expectancy of mammals is roughly proportional to their growth phase (*Second Discourse*, n. *d, p. 73, citing Buffon's *Natural History*, vol. 4, pp. 226–27), Rousseau thus explained distinct features of human ontogeny without invoking either special creation or natural discontinuities more general than climatic changes or localized events such as earthquakes and volcanic eruptions (*Second Discourse*, pt. 2, p. 152; n. *4, p. 71; n. *d, p. 73; n. *8, pp. 81–86.

28. The founding of settled homesites is described as "the epoch of a first revolution, which produced the establishment and differentiation of families and which introduced a sort of property" (*Second Discourse*, pt. 2, p. 46). The division of labor and the emergence of "property" were a "great revolution" produced by the twin discoveries of metallurgy and agriculture (*Second Discourse*, pt. 2, p. 49). The establishment of governments and laws was the result of a "social contract" invented by the rich as "the most deliberate project that ever entered the human mind" (*Second Discourse*, pt. 2, pp. 53–54). Finally, such governments—originally legitimate because based on the voluntary consent of the governed—are usurped by tyrants under whom "everything is brought back to the sole law of the stronger, and consequently to a new state of nature different from the one with which we began, in that the one was the state of nature in its purity, and this last is the fruit of an excess of corruption" (*Second Discourse*, pt. 2, p. 65).

Rousseau summarizes these transitions as a sequence of revolutionary changes, each of which ushers in an "epoch" or period of relative stability within which changes are gradual and slow. Originally humans were equal, but human history has introduced progressively greater inequalities among men. "If we follow the progress of inequality in these different revolutions, we shall find that the establishment of the law and of the right of property was the first stage, the institution of the magistracy the second, and the third and last was the changing of legitimate power into arbitrary power. So that the status of rich and poor was authorized by the first epoch, that of powerful and weak by the second, and by the third that of master and slave, which is the last degree of inequality and the limit to which all the others finally lead, until new revolutions dissolve the government altogether or bring it closer to its legitimate institution" (*Second Discourse*, pt. 2, p. 62). Although changes seem possible within each epoch, the general process is an irreversible trend toward social inequality, physical decrepitude, and moral corruption. In addition to the *Letter to Philopolis* and the *Letter to d'Alembert*; see *Social*

Contract, 3.10–11 (ed. Roger D. Masters; New York: St. Martin's Press, 1978) pp. 96–99.

29. Having asserted that humans were originally stupid and peaceful animals, Rousseau concludes that inequality is "almost null in the state of nature"; if so, he claims, "moral inequality [i.e., social or legal inequalities of wealth and power] is contrary to natural right whenever it is not combined in the same proportion with physical inequality" (*Second Discourse*, pt. 2, p. 67). Hereditary monarchy and inequality of wealth are thus illegitimate in principle. Or, to use the resounding phrase that opens Rousseau's most famous political work: "Man was [and is] born free, and everywhere he is in chains" (*Social Contract*, 1.1, p. 46). Unlike the social Darwinists of the late 19th century, therefore, Rousseau neither transferred his gradualist conception of evolution to human history, nor did he impute selective advantage to the results of the political process. See Richard Hofstadter, *Social Darwinism in American Thought* (Boston: Beacon Press, 1955) and the discussion of Sumner in the next section of this chapter. On the effects of climate in politics, see *Social Contract*, 2.8–9, pp. 70–76

30. See Roger D. Masters, *The Political Philosophy of Rousseau* (Princeton, N.J.: Princeton University Press, 1968).

31. William Graham Sumner's *What Social Classes Owe to Each Other* was first published by Harper & Brothers in 1883 and republished by Yale University Press in 1925. I have used the 1963 reprint from Caxton Printers, Caldwell, Idaho. Sumner begins from the assertion that objective knowledge of political and moral principles is possible: "God and Nature have ordained the chances and conditions of life on earth once for all. We are absolutely shut up to the need and duty, if we would learn how to live happily, of investigating the laws of Nature, and deducing the rules of right living in the world as it is" (p. 14). The passion with which many scholars attack the naturalistic fallacy may be associated with the way Sumner asserts the need to deduce rules of "ought" from the facts of "the world as it is," and does so to justify laissez-faire political preferences. But since we can find the same linkage of *Is* and *Ought* in Marx, not to mention Rousseau, Hobbes, Aristotle, and the Sophists, it is hard to attribute Sumner's conclusions to this methodological assumption.

32. "We know that men once lived on the spontaneous fruits of the earth, just as other animals do" (*What Social Classes Owe to Each Other*, p. 51). Originally human existence "was almost entirely controlled by accident," and "intelligent reflection" also emerged by "accident." Although the first steps "may have been won and lost again many times," humans developed "capital," which has ultimately made possible civilization and control of natural necessity (pp. 52–53). "Necessity—that is, the need of getting a living" (what Darwin called natural selection) is as much the cause of "industrial organization" in humans as of anything else in nature (p. 55). Such social institutions as "personal and property rights" were "gradually invented and established" (e.g., p. 89).

33. "Two men cannot eat the same loaf of bread" (*What Social Classes Owe to Each Other*, p. 60). Private property gives rise to "capital," formed by "self-denial" and giving its possessor "a great advantage" over others in "the struggle for existence" (pp. 66–67). In "a constantly advancing contest with Nature," every individual is driven to "an aggressive and conquering policy toward the limiting conditions of human life." Competition is a necessity, for "the relation of parents and children is the only case of sacrifice in Nature" (p. 64). Politics is merely an

extension of this process; for Sumner as for Marx: "The history of the human race is one long story of attempts by certain persons and classes to obtain control of the power of the State, so as to win earthly gratifications at the expense of others" (p. 88). Sumner's argument is stated in a way that combines individualistic premises with the metaphor of society as itself an organism that tends to be in equilibrium: "the characteristic of all social doctors is, that they fix their minds on some man or group of men whose case appeals to the sympathies and the imagination, and they plan remedies addressed to the particular trouble; they do not understand that all the parts of society hold together, and that forces which are set in action act and react throughout the whole organism, until an equilibrium is produced by a readjustment of all interests and rights" (p. 107).

34. Although based on "the appeal to sympathy and generosity, and to all the other noble sentiments of the human heart," Sumner dismisses "schemes of philanthropy and humanitarianism" (p. 107). "The same piece of capital cannot be used in two ways. Every bit of capital, therefore, which is given to a shiftless and inefficient member of society, who makes no return for it, is diverted from a reproductive use; but if it was put to reproductive use, it would have to be granted in wages to an efficient and productive laborer. Hence the real sufferer by that kind of benevolence which consists in an expenditure of capital to protect the good-for-nothing is the industrious laborer." (pp. 108–9). The real victim of social reform is the "industrious and sober workman": the "Forgotten Man," who, as Sumner insists, "is not infrequently a woman" (pp. 110–26).

For Sumner, as for other social Darwinists in the late 19th century, "Nature's remedies against vice are terrible. She removes the victims without pity. A drunkard in the gutter is just where he ought to be, according to the fitness and tendency of things. Nature has set up on him the process of decline and dissolution by which she removes things which have survived their usefulness." (p. 114). Sentimentalism is absurd: that "Nature's forces know no pity" is a truth in sociology as in "natural philosophy" (p. 133). Sumner also attacks the appeal to "natural rights" (or to civic or human rights) as misleading: "There can be no rights against Nature, except to get out of her whatever we can, which is only the fact of the struggle for existence stated over again" (pp. 116–17). On this ground, Sumner attacks not only social welfare legislation, but "internal improvements" to rivers and canals, the protective tariff, and any scheme of governmental "paternalism" to benefit the rich. Indeed "all prohibitory, sumptuary, and moral legislation" of whatever kind represents the same "fallacy" of interfering with "liberty" (p. 115).

35. See Petr Alexseevich Kropotkin, *Mutual Aid* (New York: N.Y.U. Press, 1974). Moreover, it is a delicious irony that Sumner himself makes a strong argument that his theory, like all science, is "value free": "political economy . . . treats of the laws of the material welfare of human societies. It is, therefore, only one science among all the sciences which inform us about the laws and conditions of our life on earth. There is no injunction, no 'ought' in political economy at all. It does not assume to tell man what he ought to do, any more than chemistry tells us that we ought to mix things, or mathematics that we ought to solve equations. It only gives one element necessary to an intelligent decision, and in every practical and concrete case the responsibility of deciding what to do rests on the man who has to act" (*What Social Classes Owe to Each Other*, pp. 134–35). Like many 20th-century postmodernists, Sumner was hostile to the self-interest of the rich (pp. 89–90), concerned about the interests of women (chap. 10), and emphasized

that "rights must be equal" in order to "guarantee mutually the chance to earn, to possess, to learn, to marry, etc." (p. 141). As is often the case in the history of thought, stereotypes obscure as much as they reveal.

36. See Richard Lewontin, Steven Rose, and Leon Kamin, *Not in Our Genes* (Boston: Pantheon, 1984); Arther Caplan, *The Sociobiology Debate* (Garden City, N.Y.: Doubleday Anchor, 1979). Marx had, on seeing Darwin's *Origin of Species*, commented on this point, accusing Darwin of transforming Hobbes's "war of all against all" into a universal principle of evolution under the label "survival of the fittest." Many critics have also asserted that sociobiology reduces all human behavior to genetic determinism. But this latter charge is so manifestly contrary to the theory that it need not be explored in detail here. As Richard Alexander noted, sociobiological theories would be unaffected were it to be shown that human behavior was entirely due to individual learning governed by the laws of behaviorist psychology. See Richard Alexander, *Darwinism and Human Affairs* (Seattle: University of Washington Press, 1979); Roger D. Masters, "The Value—and Limits— of Sociobiology," in *Sociobiology and Human Politics*, ed. Elliott White (Lexington, Mass.: Lexington Books, 1981).

37. The evidence summarized here is described in more detail in Roger D. Masters, "Is Sociobiology Reactionary? The Political Implications of Inclusive Fitness Theory." *Quarterly Review of Biology* 57 (1982): 275–92. William Charlesworth has wisely concluded that the question concerns the rhetorical use of theories rather than the theories themselves:

Speaking of rhetoric, there should be an editorial rule that sentences associating sociobiology with effort to justify slavery, imperialism, racism, genocide, and to oppose equal rights should always appear next to sentences associating environmentalist/learning theory with effort to justify propaganda, psychological terror, false advertisement, public indoctrination of hatred of foreigners, class enemies, minority groups, and so on and so on. Juxtaposing sociology and learning theory in this manner ought to show how unproductive it is to claim through innuendo or otherwise that science will lead to pseudoscience, will lead to man's inhumanity to man: ergo no science. ("Comments on S. L. Washburn's Review of Kenneth Bock's 'Human Nature History,' *Human Ethology Newsletter*, September 22, 1981, p. 22).

38. In the *Republic*, after Socrates tries to refute Thrasymachus's argument that justice is the "interest of the stronger," Glaucon restates the views of Thrasymachus and the Sophists more generally, as follows: "They say that doing injustice is naturally good, and suffering injustice bad, but that the bad in suffering injustice far exceeds the good in doing it; so that, when they do injustice to one another and suffer it and taste of both, it seems profitable—to those who are not able to escape the one and choose the other—to set down a compact among themselves neither to do injustice nor to suffer it. And from there they began to set down their own laws and compacts and to name what the law commands lawful and just" (*Republic*, 2.358e–359a; pp. 36–37). For Antiphon, see Ernest Barker, *Greek Political Thought* (New York: Barnes and Noble, 1968), or Louis Gervet, ed., *Antiphon: Discours suivi des fragments d'Antiphon le Sophiste* (Paris: Les Belles Lettres, 1923). On the entire tradition, see Masters, *Nature of Politics*, chap. 1.

39. Taijfel and Peters, "Hobbes and Hull." Cf. Eric Alden Smith and Bruce Winterhalder, *Evolutionary Ecology and Human Behavior* (New York: Aldine de Gruyter, 1992), pp. 39–47.

40. *Politics*, esp. 1.1–10. On the pre-Socratic view and the reaction to it by Plato and Aristotle, and the Socratic tradition, see Roger D. Masters, "Classical Political Philosophy and Contemporary Biology," in *Politikos: Selected Papers of the American Conference of Greek Political Thought*, ed. Kent Moors, (Pittsburgh: Duquesne University Press, 1989), vol. 1, pp. 1–44.

41. See Kropotkin, *Mutual Aid*; Adolph Portmann, *Animals as Social Beings* (London: Hutchinson, 1961); David Sloan Wilson, *The Natural Selection of Populations and Communities* (Menlo Park, Calif.: Benjamin/Cummings, 1979). It is only fair to note that some evolutionary biologists who stress natural selection at the individual level have also recognized that, under specified conditions, group selection also operates: e.g., Edward O. Wilson, *Sociobiology* (Cambridge, Mass.: Harvard University Press, 1975), chap. 5.

42. Larry Laudan, *Science and Relativism: Some Key Controversies in the Philosophy of Science* (Chicago: University of Chicago Press, 1990), p. 128. Cf. George Johnson, *In the Palaces of Memory* (New York: Vintage, 1992), pp. 224–26.

43. See, inter alia, Claude Lévi-Strauss, *Anthropologie structurale* (Paris: Plon, 1958), *La pensée sauvage* (Paris: Plon, 1962), and *Mythologiques III: L'origine des manières de table* (Paris: Plon, 1968).

44. Those who continue to claim that the truth of science is contingent on historical values because all human thought is situated in time and space must admit a paradoxical feature of their argument. According to modern physics, all events are so situated; to some degree, it is the relativist or intuitive thinker who demands a standard of knowledge that is more absolute than the scientist or traditional philosopher (see Laudan, *Science and Relativism*, p. 108 et passim). And if all thought is conditioned by the circumstances of its origin, surely this insight must apply to the relativist position as well. The aggressive critic of relativism might then ask what motivates those who deny all scientific knowledge? Relativists have often attacked scientists for supporting the status quo, Western civilization, the capitalist order, male chauvinism, and other sins. The same technique could be used to assert that relativists and intuitivists attack science as a means of denying any form of knowledge that might be used as a constraint on their own individual choices. Whether as the basis for justifying individual hedonism or cultural diversity, this rhetoric could be turned against the critics of science. Such debates miss the point, however. Relativism, like methodological individualism or historical gradualism, has been associated with the most diverse theoretical positions over the history of Western thought.

7. Scientific Value Relativism

1. For the general public in the West, one of the principal challenges to the modern scientific and technological worldview has recently come from the concern for environmental protection; it is symbolic that supporters of this movement are often called "ecologists" (taking the name of a scientific field as the label for a social concern) and typically support scientific analyses of environmental degradation rather than openly attacking science itself. Among major Western philosophers, there is a similar tendency to accept the principles of modern science and to focus instead on the way scientific knowledge is used. The major exception is probably Rousseau's *Discourse on the Sciences and Arts*, or *First Discourse* (*Collected Writings*

of Rousseau, ed. Roger D. Masters and Christopher Kelly [Hanover, N.H.: University Press of New England, 1992], vol. 2, pp. 1–22)—an exception that proves the rule, since Rousseau himself explicitly appeals to ancient philosophy as the basis of his criticism of modern science; see Roger D. Masters, *The Political Philosophy of Rousseau* (Princeton, N.J.: Princeton University Press, 1968). In more recent times, thinkers like C. S. Lewis have presented a similar critique; see *The Abolition of Man* (New York: Collier, 1962). In any case, direct attacks on the natural sciences by intellectuals have not generated as deep a popular echo in the West (with the possible exception of hostility to nuclear technology) as religious rejection of the scientific project continues to engender in some non-Western cultures.

2. See chap. 1. For a survey of the issues of science and public policy, see Michael Kraft and Norman Vig. eds., *Technology and Politics* (Durham, N.C.: Duke University Press, 1988). As my friend Michael Platt pointed out a complete treatment of the nature and implications of modern science would require a consideration of the work of Michael Polanyi, Alfred North Whitehead, E. O. Burt, Herbert Butterfield, Kurt Riezler, Charles de Koninck, Vincent Smith, M. B. Foster, Jacob Klein, and many others. In the pages that follow, however, my intention is to direct the reader toward a more complete understanding of issues that have generally been taken for granted, rather than to pretend to offer a definitive answer.

3. Cf. Bacon, *New Organon*, 1.80 (ed. Sidney Warshaft [New York: Odyssey, 1965], p. 353); Hobbes, *Leviathan*, ed. Herbert W. Schneider (New York: Library of Liberal Arts, 1958), 1.9, pp. 75–77; Rousseau, *First Discourse* (*Collected Writings*, vol. 2, p. 7).

4. Mihaly Csikszentmihalyi, "Review of *Childhood* by Melvin Konner," *New York Times Book Review*, November 10, 1991, p. 16.

5. See John Locke, *Two Treatises of Civil Government*, ed. Thomas I. Cook (New York: Hafner, 1956). The classic statement of Locke's impact of American political traditions is probably still Louis Hartz, *The Liberal Tradition in America* (New York: Harcourt Brace, 1955).

6. John Locke, *An Essay Concerning Human Understanding*, 1.2.5; ed. Alexander Campbell Fraser (New York: Dover, 1959), vol. 1, p. 69. All references in the text are to this edition. Note that, for Locke, the problem of knowledge is inseparable from the question of moral judgment and obligation. Although contemporary philosphers often separate the study of ethics and morality from epistemology, many major thinkers have rejected this typically modern specialization. Locke's view that the philosophy of knowledge is directly associated with moral questions is shared, for example, by Plato, Hume, and Kant, not to mention Aristotle, Machiavelli, Rousseau, Hegel, or Nietzsche.

7. While contrary to many scholarly interpretations of Locke, a close examination of his work shows the marked influence of Hobbesian premises (if not always leading to the same conclusions). For a fuller statement of the evidence, see Richard H. Cox, *Locke on War and Peace* (Oxford: Clarendon Press, 1960), and Roger D. Masters, "Hobbes and Locke," in *Comparing Political Thinkers*, ed. Ross Fitzgerald (Sydney and New York: Pergamon Press, 1980), pp. 116–40.

8. "He that will carefully peruse the history of mankind, and look abroad into the several tribes of men, and with indifferency survey their actions, will be able to satisfy himself that there is scarce that principle of morality to be named, or rule of virtue to be thought on, (those only excepted that are absolutely necessary to hold society together, which commonly too are neglected betwixt distinct societies,)

which is not, somewhere or other, slighted and condemned by the general fashion of whole societies of men, governed by practical opinions and rules of living quite opposite to others" (*Essay*, 1.2.10; vol. 1, p. 74). Even the most basic principle of ethics, the Golden Rule, is not reliable: "The great principle of morality, 'To do as one would be done to,' is more commended than practised" (1.2.7; vol. 1, p. 71). And this follows not merely because humans lack knowledge, but—perhaps even more important—because the principle of the conscience used to buttress the traditional natural law teaching since Aquinas is equally unreliable; as Locke asserts, "I cannot see how any men should ever transgress these moral rules, with confidence and serenity, were they innate, and stamped upon their minds" 1.2.9; vol. 1, p. 72.

9. Locke's argument rests on the premise that "no practical rule which is anywhere universally, and with public approbation or allowance, transgressed, can be supposed innate" (*Essay*, 1.2.11; vol. 1, p. 75). That is, Locke uses the variety of social customs as what would today be called a "natural experiment": if all members of a culture have adopted a practice, it cannot be contrary to instinct. Locke's argument is ingenious, because it contains an implicitly statistical view of the way to test the argument for innate ideas of morality.

10. Plato, *Republic*, esp. 4. 428e, 433a–343b, 444d (ed. Allan Bloom; New York: Basic Books, 1968, pp. 110, 111–12, 124); Aristotle, *Politics*, esp. 1.2.1253a–b; Leo Strauss, *The City and Man* (Chicago: Rand McNally, 1964), as well as *Natural Right and History* (Chicago: University of Chicago Press, 1953).

11. "Principles of actions indeed there are lodged in men's appetites; but these are so far from being innate moral principles, that if they were left to their full swing they would carry men to the overturning of all morality" (*Essay*, 1.2.13; vol. 1, p. 77). "There is a great deal of difference between an innate law, and a law of nature; between something imprinted on our minds in their very original, and something that we, being ignorant of, may attain to the knowledge of, by the use and due application of our natural faculties. And I think they equally forsake the truth who, running into contrary extremes, either affirm an innate law, or deny that there is a law knowable by the light of nature, i.e., without the help of positive revelation" (*Essay*, 1.2.13; vol. 1, p. 78). Cf. Bacon's definition of "laws of nature" (*New Organon*, 2.17 [ed. Warshaft, p. 388]).

12. To use the blunt language of Hobbes, "there is no such *finis ultimus, utmost aim* or *summum bonum, highest good*, such as is spoken of in the books of the ancient philosophers" (*Leviathan*, 1.11, ed. Herbert W. Schneider (New York: Library of Liberal Arts, 1958), p. 86). Cf. Locke's definition of ethics in *Essay*, 4.21.1; vol. 2, p. 460.

13. John Locke, *Second Treatise of Civil Government*. Locke's political ideas are sufficiently well known and so directly associated with contemporary American practice that I need not discuss them at length; it should suffice to have shown how these political principles are derived in important respects from the epistemology set forth in Locke's *Essay*.

14. *Natural Right and History*, pp. 7–8.

15. E.g., Michael Ruse, *Taking Darwin Seriously* (Oxford: Blackwell, 1986), p. 87.

16. David Hume, *A Treatise of Human Nature*, 1.3.2 (Garden City, N.Y.: Doubleday, Anchor, 1961), p. 66. As originally stated (1.1.1; p. 6): Hume asserts *"that all our simple ideas in their first appearance, are derived from simple impressions, which are correspondent to them, and which they exactly represent"* (italics

in both citations in the original). Note that Hume's apparently decisive evidence for this principle is an empirical observation: "wherever, by any accident, the faculties which give rise to any impressions are obstructed in their operations, as when one is born blind or deaf, not only the impressions are lost, but also their correspondent ideas; so that there never appear in the mind the least trace of either of them" (1.1.1; p. 5). Sensory deprivation experiments confirm that a newborn animal artificially prevented from sensing a specific stimulus cue (such as a color, curved or straight lines, or phonemic contrasts) will mature without the neuronal structures and pathways which normally process that sensory cue; in humans, however, such studies do not show the corresponding inability to *think* of the cue. On these deficits, which were of great interest to Locke and his successors in the enlightenment, see Jacques Ninio, *L'empreinte des sens* (Paris: Odile Jacob, 1989), chap. 2.

17. "The only connection or relation of objects, which can lead us beyond the immediate impressions of our memory and senses, is that of cause and effect . . . The idea of cause and effect is derived from *experience*, which informs us, that such particular objects, in all past instances, have been constantly conjoined with each other; and as an object similar to one of these is supposed to be immediately present in its impression, we thence presume on the existence of one similar to its usual attendant" (Hume, *Treatise*, 1.3.6; p. 82).

18. "Thus, not only our reason fails us in the discovery of the *ultimate connection* of causes and effects, but even after experience has informed us of their *constant conjunction*, it is impossible for us to satisfy ourselves by our reason, why we should extend that experience beyond those particular instances which have fallen under our observation" (ibid, p. 84.) Cf. chap. 3 above, esp. Fig. 3.1.

19. As the context makes clear, however, for Hume this is a defect of the "vulgar systems of morality" seeking to derive duty or virtue from *reason*; this error can be avoided by deriving moral distinctions from a "moral sense" rooted in nature, and especially in the natural sentiment of "sympathy" (ibid., 3.1.2; pp. 424ff.). Since these natural sentiments are discovered by scientific evidence, Hume's procedure differs sharply from the contemporary authors discussed in the next section.

20. Hume's view is clearly more defensible than the argument that scientific prediction is (at least in principle) a rigorously deterministic "covering law" from which events can be deduced with logical certainty: cf. Carl G. Hempel and Paul Oppenheim, "The Logic of Explanation," *Philosophy of Science* (15 (1948):135–74. More specifically Hume's complex view of additive or "compounded" probabilities mirrors such contemporary statistical analyses as multiple regression and analysis of variance. Hume, *Treatise*, 1.3.12: "[W]e may establish it as a certain maxim, that in all moral as well as natural phenomena, wherever any cause consists of a number of parts, and the effect increases or diminishes, according to the variation of that number, the effect, properly speaking, is a compounded one, and arises from the union of the several effects, that proceed from each part of the cause. . . . The absence or presence of a part of the cause is attended with that of a proportionable part of the effect" (p. 125). On multiple regression equations (the modern mathematical equivalent of Hume's explanation) as the model for the operation of the brain, compare A. P. Georgopoulos, A. B. Schwartz, and R. E. Kettner, "Neuronal Population Coding of Movement Direction," *Science* 233 (1986):1416–19.

21. "The several instances of resembling conjunctions lead us into the notion of power and necessity. . . . Necessity, then, is the effect of this observation, and is nothing but an internal impression of the mind, or a determination to carry our

thoughts from one object to another" (Hume, *Treatise*, 1.3.14; (pp. 151–52). In this context, Hume explicitly applies this notion to both arithmetic and geometry ("the necessity, which makes two times two equal to four, or three angles of a triangle equal to two right ones").

22. See Alaisdair MacIntyre, "Hume on 'Is' and 'Ought,'" *Philosophical Review* 68 (1959):451–678; Robert McShea, *Morality and Human Nature* (Philadelphia: Temple University Press, 1990).

23. Cf. May Broadbeck, *Readings in the Philosophy of the Social Sciences* (New York: Macmillan, 1968); Strauss, *Natural Right and History*. In philosophy, the same issue is posed in terms of the "naturalistic fallacy," for which the most frequently cited source is probably G. E. Moore's *Principia Ethica* (Cambridge: Cambridge University Press, 1903); see Ruse, *Taking Darwin Seriously*, pp. 88–90.

24. Locke begins by a categorization of these branches in a descriptive manner: "All that can fall within the compass of human understanding, being either, *First*, the nature of things, as they are in themselves, their relations, and their manner of operation: or, *Secondly*, that which man himself ought to do, as a rational and voluntary agent, for the attainment of any end, especially happiness; or, *Thirdly*, the ways and means whereby the knowledge of both the one and the other of these is attained and communicated; I think science may be divided properly into these three sorts" (*Essay*, 4.21.1; vol. 2, p. 460). Then Locke provides each with its name: first, "the knowledge of things, as they are in their own proper beings, then constitution, properties, and operations; whereby I mean not only matter and body, but spirits also. . . . This, in a little more enlarged sense, of the word, I call PHYSIKE, or *natural philosophy*. . . . Secondly, PRAKTIKE, the skill of right applying our own powers and actions, for the attainment of things good and useful. The most considerable under this head is *ethics*. . . . Thirdly, the third branch may be called SEMEIOUTIKE, or *the doctrine of signs*; the most usual whereof being words, it is aptly enough termed also LOGIKE, *logic*" (Locke, *Essay*, 4.21.2–4; vol. 2, p. 461). The end of physics is "bare speculative truth"; of ethics, "not bare speculation and the knowledge of truth, but right, and a conduct suitable to it"; in contrast, "the business" of semiotics or logic "is to consider the nature of signs, the mind makes use of for the understanding of things, or conveying its knowledge to others" (ibid). Insofar as the study of literature and the arts is today focused on semiotics rather than claims to substantive truth, Locke's categories seem to parallel the contemporary division between natural sciences, social sciences and humanities dominating our universities (cf. the following note).

25. Thus, one philosopher of science begins a critique of political science, treated as the epitome of the social or "practical" sciences, by asserting that "the basic judgments of human beings may be divided into three fundamentally different categories: (1) logical judgments, (2) factual judgments, and (3) value judgments" (John Kemeny, "A Philosopher Looks at Political Science," *Conflict Resolution* Vol. 4 [1960]:p. 292). On the distinction between logic and either fact (or "laws of nature") or value, see also Arnold Brecht, *Political Theory* (Princeton, N.J.: Princeton University Press, 1968), pp. 66–68.

26. Johann Hjort, *The Human Value of Biology* (Cambridge, Mass.: Harvard University Press, 1938); Gunther Stent, ed., *Morality as a Biological Phenomenon* (Berkeley: University of California Press, 1979).

27. "If we attend to the experience of men's conduct, we meet frequent and, as we admit ourselves, just complaints that there is not to be found a single certain

example of the disposition to act from pure duty. Although many things are done in conformity to what duty prescribes, it is nevertheless always doubtful whether they are done strictly out of duty" (Immanuel Kant, *Metaphysical Foundations of Morals*, in *The Philosophy of Kant*, ed. Carl J. Friedrich [New York: Modern Library, 1964], p. 154).

28. Ibid., p. 174.

29. Probably the most striking example is John Rawls, *A Theory of Justice* (Cambridge, Mass.: Harvard University Press, 1971). For a valuable critique of contemporary neo-Kantian ethical thought from a naturalistic perspective, see Ian Shapiro's *Political Criticism* (Berkeley: University of California Press, 1990). See also Martha C. Nussbaum, *The Fragility of Goodness: Luck and Ethics in Greek Tragedy and Philosophy* (Cambridge: Cambridge University Press, 1986).

30. Perhaps the classic example of this criticism as it relates to "science and historical knowledge" in general (and Marxism in particular) is Isaiah Berlin, *Historical Inevitability* (London: Oxford University Press, 1954). Berlin does not seem to notice that his objection takes the form of a general rule or scientific "law" of human affairs: "no sooner do we acquire adequate insight into the 'inexorable' and 'inevitable' parts played by all things animate and inanimate in the cosmic process, than we are freed from the sense of personal endeavor. Our sense of guilt and of sin, our pangs of remorse and self-condemnation, are automatically dissolved" (*Historical Inevitability*, as reprinted in Leonard I. Krimerman, *The Nature and Scope of Social Science* [New York: Appleton Century Crofts, 1969], p. 685). Ironically, Berlin's scientific prediction about the way humans respond to determinist theories of history is contradicted by the empirical evidence of both Calvinist responses to the doctrine of predestination and the behavior of Stalinists who believed in the inevitability of dialectical materialism. For Skinner's analysis of values, see B. F. Skinner, *Beyond Freedom and Dignity* (New York: Knopf, 1971). In this case, the irony is not that Skinner's argument is a determinist account of human nature which undermines human ability to control events, but that his science—based on conditioning and individual experience—followed Lockean individualism at a time when the biological sciences were establishing the basis of a far more powerful and important theory of social behavior, within which Skinnerean condition has a limited albeit demonstrable role.

31. See also T. D. Weldon, *The Vocabulary of Politics* (Harmondsworth, England: Penguin, 1953), as well as the selections in Broadbeck, *Readings in the Philosophy of Social Science*. My discussion in the following section has benefited enormously from years of dialogue with my colleague Denis G. Sullivan, with whom I taught a course titled "The Nature of Political Inquiry" for over a decade. While he may not agree with my conclusions, I have been especially influenced by his presentation of the "covering law" model of scientific explanation (see n. 50) and its use to show that all values are ultimately derived from facts or scientific generalizations presumed to be factually true.

32. "Deductive analytic logic . . . can add nothing to the meaning of propositions; it can merely make explicit what is implied in that meaning. Inferences of what 'ought' to be, therefore, can never be derived deductively (analytically) from premises whose meaning is limited to what 'is'; they can be correctly made only from statements that have an Ought-meaning, at least in the major premise. . . . In logic there is, as some have expressed it, an 'unbridgeable gulf' between Is and Ought" (Brecht, *Political Theory*, p. 126–27). At this point, Brecht cites Max

Weber, Emil Lask, and Hans Kelson—and, as the citations in my notes will confirm, this position reflects the tradition that dominates social science in American universities generally.

33. In technical terms, this has been described as the problem of infinite regress or reflexivity. On the question of whether reflexivity invalidates all social scientific theory, see Roger C. Buck, "Reflexive Predictions," *Philosophy of Science* 30 (1963). Recently, similar issues have been considered in terms of artificial intelligence, where they take the form of what is called the "stopping problem." See Roger Penrose, *The Emperor's New Mind* (New York: Oxford University Press, 1989).

34. That pure mathematics entails ethical choice in the deepest sense of the term is amply confirmed by David R. Lachterman, *The Ethics of Geometry: A Genealogy of Modernity* (New York: Routledge, 1989), esp. chap. 1.

35. The practice of peer review is, of course, a direct reflection of this premise, reflecting an extension of the assumptions of a republican or democratic form of government to the internal decisions of a scientific discipline. On the problems arising from this mode of assessment, see Dominic V. Cicchetti, "The Reliability of Peer Review for Manuscript and Grant Submissions: A Cross-Disciplinary Investigation," *Behavioral and Brain Sciences* 14 (1991):119–86.

36. For example, Kemeny solves the problem of premises in a simple manner: "It is common advice given to a student in college to copy the activities of his more successful roommate. Similarly, the social scientist is well advised to learn lessons from the success of the physical scientist" (John Kemeny, "A Philosopher Looks," p. 297.

37. Cf. Michel Foucault, *The Order of Things* (New York: Viking, 1970); David Berlinski, *Black Mischief: Language, Life, Logic, Luck*, 2nd ed. (Boston: Harcourt, Brace Jovanovich, 1988); Paul Feyerabend, *Against Method*, (London: NLB; Atlantic Highlands, N.J.: Humanities, 1975). On the origins of this movement in Nietzsche's thought, see Allan Bloom, *The Closing of the American Mind* (New York: Simon and Schuster, 1986), pt. 2; and Leslie Paul Thiele, "The Agony of Politics: The Nietzschean Roots of Foucault's Thought," *American Political Science Review* 84 (1990):907–25.

38. To be consistent, scholars holding this position should accept the teaching of creation science on an equal footing with evolutionary biology (cf. Brecht, *Political Theory*, pp. 87–88, in part cited in n. 44 to chap. 8).

39. As has been noted, writers who affirm the gulf between fact and value often violate their own strictures when presenting scientific method. To cite Kemeny again, "If a science is to develop, it must be allowed a maximum amount of freedom in the subject matter it studies and the conclusions it is to reach. . . . The fear in allowing a department of political science to explore all existing, and many unheard-of, political systems lies in the concern of a government that these scientific theories may prove one's own form of government not ideal. It *must*, therefore, be understood that a political scientist, as a scientist, can never come to such a conclusion. But neither can he come to the conclusion that one's present form of government is ideal. These judgments must be left to the political moralists" (Kemeny, "A Philosopher Looks," p. 302; italics in the original). Note that the political vulnerability of science—the issue confronting Socrates (see Plato's *Apology* as well as Aristophanes's *Clouds*)—apparently applies to modern science as well as to that of the ancients. If the efficacy of modern science is its principal recommendation, this measure of utility does not seem to lie primarily in resolving the primordial tension

between scientific or philosophic knowledge and the popular opinions reflected in politics.

40. For instance, John Kemeny rejected Julian Huxley's claim that "scientific facts can imply moral judgments" with the following revealing statement: "Fortunately, the position represented by Huxley and others, known to philosophers as the *naturalistic fallacy*, is now generally rejected. . . . Rather than get involved in a lengthy philosophical discussion at this point, I shall take for granted a position now shared by the majority of philosophers of science" (Kemeny, "A Philosopher Looks," p. 293. It is hard to imagine a more "democratic" criterion of scientific judgment than the "majority" of those in the appropriate discipline. The evidence from the history of science shows the difficulty implicit in this view (e.g., Thomas Kuhn's *The Structure of Scientific Revolutions* [Chicago: University of Chicago Press, 1970] and the many works spawned to attack or support his interpretation).

41. See, for example, Rousseau's *Emile*, ed. Allan Bloom (New York: Basic Books, 1979), p. 458: "It is necessary to know what ought to be in order to judge soundly about what is"; Hegel, Preface, *Philosophy of Right*, ed. T. M. Knox (New York: Oxford University Press, 1967), pp. 10–13; Marx and Engels, *German Ideology*, pp. 37, 42–48, 106–7.

42. "[S]ince vice and virtue are not discoverable merely by reason, or the comparison of ideas, it must be by means of some impression, or sentiment they occasion, that we are able to mark the difference betwixt them. . . . Morality, therefore, is more properly felt than judged of" (*Treatise*, 3.1.2; p. 424). It cannot be pretended that Hume was unaware of his critique of causality when he made this assertion, for the sentence continues: "though this feeling or sentiment is commonly so soft and gentle that we are apt to confound it with an idea, according to our common custom of taking all things for the same which have any near resemblance to each other" (*Treatise*, 3.1.2; p. 424).

43. The passage is of sufficient importance to be cited at length:

should it be asked, whether we ought to search for these principles in *nature*, or whether we must look for them in some other origin? I would reply, that our answer to this question depends on the definition of the word Nature, than which there is none more ambiguous and equivocal. If *nature* be opposed to miracles, not only the distinction betwixt vice and virtue is natural, but also every event which has ever happened in the world, *excepting those miracles on which our religion is founded*. . . . But *nature* may also be opposed to rare and unusual; and in this sense of the word, which is the common one, there may often arise disputes concerning what is natural or unnatural; and one may in general affirm, that we are not possessed of any very precise standard by which these disputes can be decided. Frequent and rare depend upon the number of examples we have observed; and as this number may gradually increase or diminish, it will be impossible to fix any exact boundaries betwixt them. We may only affirm on this head, that if ever there was anything which could be called natural in this sense, the sentiments of morality certainly may; since there never was any nation of the world, nor any single person in any nation, who was utterly deprived of them and who never, in any instance, showed the least approbation or dislike of manners. . . .

But *nature* may also be opposed to artifice, as well as to what is rare and unusual; and in this sense it may be disputed whether the notions of virtue be natural or not . . . Perhaps it will appear afterwards that our sense of some virtues is artificial, and that of others natural. The discussion of this question will be more proper, when we enter upon an exact detail of each particular vice and virtue. (*Treatise*, 3.1.2; pp. 427–28)

44. On the importance of this distinction between artificial and natural virtues, see J. B. Schneewind, "The Misfortunes of Virtue," *Ethics* 101 (1990):42–63, esp. 50–54.

45. Hume's argument rests on the principle that the praise of virtue (or blame of vice) is directed to the motive for an action rather than the act per se; otherwise, as Hume observes, we would hold the acorn responsible for parricide if it grows so tall that it overshadows and kills the oak from which it came (*Treatise*, 3.1.1; p. 421). It follows "*that no action can be virtuous, or morally good, unless there be in human nature some motive to produce it distinct from the sense of its morality*" (3.2.1; p. 431). There is always, therefore, a motive—be it "natural" (as love of self or of children) or "artificial" and learned (devotion to the public interest) which lies at the root of the "sense of justice" or other virtues. Virtues themselves can be said to be natural if both the motive and the actions corresponding to them are natural, whereas virtues are artificial if the motive for them can only arise from education and human conventions. Because "there is no such passion in human minds as the love of mankind, merely as such, independent of personal qualities, of services, or of relation to ourself" (3.2.1; p. 433–34), "the sense of justice and injustice is not derived from nature, but arises artificially, *though necessarily*, from education and human conventions" (3.2.1; p. 435; italics added). The conjunction of the words *artificially* and *necessarily* marks the difference between Hume and recent thinkers who claim to derive the fact—value dichotomy from his work.

46. In contemporary terms, choice theory and sociobiology make a distinction between self-interest and altruism similar to the one presented by Hume. See, among other work, Howard Margolis, *Selfishness, Altruism, and Rationality* (Cambridge: Cambridge University Press, 1982); Robert Frank, *Passions within Reason* (New York: Norton, 1988); Masters, *The Nature of Politics*, esp. chap. 5.

47. Hume states his view with precision:
To avoid giving offence, I must here observe, that when I deny justice to be a natural virtue, I make use of the word *natural*, only as opposed to *artificial*. In another sense of the word, as no principle of the human mind is more natural than a sense of virtue, so no virtue is more natural than justice. Mankind is an inventive species; and where an invention is obvious and absolutely necessary, it may as properly be said to be natural as anything that proceeds immediately from original principles, without the intervention of thought or reflection. Though the rules of justice be *artificial*, they are not *arbitrary*. Nor is the expression improper to call them *Laws of Nature*; if by natural we understand what is common to any species, or even if we confine it to mean what is inseparable from the species. (*Treatise*, 3.2.1; p. 436, italics in original).
For the neo-Darwinian view, see the essays in Roger D. Masters and Margaret Gruter, eds., *The Sense of Justice: Biological Foundations of Law* (Newbury Park, Calif.: Sage, 1992).

48. See MacIntyre, "Hume on 'Is' and 'Ought' "; McShea, *Morality and Human Nature*; and Ruse, *Taking Darwin Seriously*, esp. p. 267: "Hume's 'sentimentalist' theory of morality is precisely that which one would expect as the precursor of Darwinism." I am indebted to Larry Arnhart for the reminder of the importance of this interpretation of Hume.

49. McShea, *Morality and Human Nature*, p. 226.

50. That feelings of "love" toward children, nephews, cousins, and strangers should decline as a function of the closeness of kinship (Hume, *Treatise*, 3.2.1, p. 436, cited above) is precisely what is predicted by William D. Hamilton's principle of "inclusive fitness" and demonstrated by contemporary sociobiology (William D. Hamilton, "The Genetical Theory of Social Behavior," *Journal of Theoretical Biology* 7 [1964]: 1–51; David Barash, *Sociobiology and Behavior*, 2nd ed. (New York: Elsevier, 1982); and the references in n. 46, above. It follows, moreover, that cooperation with strangers and obedience to the laws of the centralized state is in a sense "unnatural" (Masters, *The Nature of Politics*, chap. 5).

8. From the Naturalistic Fallacy to Naturalism

1. Manifest by the optimistic assertion that this is the "best of all possible worlds" in Pope's *Essay on Man*, in the Christian tradition this view is exemplified by Paul's argument in Romans, esp. 13:1. Pope's version of this view was, of course, ridiculed by Voltaire in *Candide* and the *Poem on the Disaster of Lisbon*.

2. Intuitive judgments, themselves reflecting habit or custom more than reason, all too easily cover themselves with the cloak of universality, absolute certainty, and self-evident truth. For example, see again the list of supposedly "universal" elements of the sense of justice that are discussed by Arnold Brecht, *Political Theory*, (Princeton, N.J.: Princeton University Press, 1968), chap. 10.

3. On one level, Locke, Hume, Descartes, and the modern philosophic traditions they established erroneously equate knowledge with the experience of the isolated individual. For a number of contemporary scholars, this approach needs to be replaced by an "evolutionary epistemology" which considers knowledge in terms of the acquisition and transmission of information conducive to the survival of any species. One of the best-known exponents of this view is Donald T. Campbell, e.g., "Evolutionary Epistemology," in P. A. Schilpp, ed., *The Philosophy of Karl Popper* (LaSalle, Ill.: Open Court, 1974), pp. 413–63. See also James H. Fetzer, ed. *Sociobiology and Epistemology* (Dordrecht: Reidel, 1985). While mere survival is inadequate as evidence of the truth of an organism's perception of the world, moreover, James Barham shows that this assumption is not necessary and, indeed, undermines a strictly evolutionary approach to knowledge (Barham, "A Poincaréan Approach to Evolutionary Epistemology," *Journal of Social and Biological Structures* 13 [1991]: 193–258). Barham's approach is particularly valuable because it treats the adaptation of organisms from the perspective of an "oscillator" or "strange attractor" that is similar to the mathematical models in chaos theory discussed in chap. 7. In other words, Barham's approach to evolutionary epistemology does not assume the kind of linear causality and logic that has been challenged in contemporary physics and mathematics (cf. chap. 2 above). While I will return to the social character of knowledge implied by these developments, the Lockean tradition also needs to be tested against empirical evidence concerning individual perception and judgment.

4. Locke puts it as follows: "First, our senses, conversant about particular sensible objects, do convey into the mind several distinct perceptions of things, according to those various ways wherein those objects do affect them. And thus we come by those *ideas*, we have of *yellow, white, heat, cold, soft, hard, bitter, sweet*,

and all those which we call sensible qualities; which when I say the senses convey into the mind, I mean, they from external objects convey into the mind what produces there those perceptions. This great source of most of the ideas we have, depending wholly upon our senses, and derived by them to the understanding, I call SENSATION" (*Essay*, 2.1.3; vol. 1, p. 122–23).

5. "Secondly, the other fountain from which experience furnisheth the understanding with ideas is,—the perception of the operations of our own mind within us, as it is employed about the ideas it has got;—which operations, when the soul comes to reflect on and consider, do furnish the understanding with another set of ideas, which could not be had from things without. And such are *perception, thinking, doubting, believing, reasoning, knowing, willing*, and all the different actings of our own minds;—which we being conscious of, and observing in ourselves, do from these receive into our understandings as distinct ideas as we do from bodies affecting our senses" (*Essay*, 2.1.3; vol. 1, p. 123).

6. Compare Locke's view of the status of material things according to the labor theory of value: *Second Treatise of Government*, chap. 5.

7. Robert Sprinkle, "Political Ethics and the Life Sciences" (Ph.D. thesis, Princeton University, 1990).

8. For an introduction, see Michael Gazzaniga, *The Social Brain* (New York: Basic Books, 1985). If by "mind" Locke meant "consciousness," then his distinction would still be inconsistent with neuroscientific evidence, since the active properties that Locke associates only with "reflection" are essential to the entire process of processing sensory input. See also Michael Gazzaniga, *Mind Matters* (Boston: Houghton Mifflin, 1988). Actually, Locke seems to associate "consciousness" with "soul," distinguishing between "mind" and "soul" much as contemporaries distinguish the central nervous system and cognition. Although Locke seems to use the word *idea* to include any abstract representation of a thing, one cannot salvage his theory by equating ideas and words because the recognition and use of words reflects the conjunction of multiple innate processing capacities and individual experience. Cf. Steven E. Peterson, Peter T. Fox, Abraham Z. Snyder, and Marcus E. Raichle, "Activation of Extrastriate and Frontal Cortical Areas by Visual Words and Word-Like Stimuli," *Science* 249 (1990): 1041–44.

9. *Essay*, 2.1.6–8; vol. 1, pp. 125–27. Cf. Aristotle, *Generation of Animals*, 1.28 (722a).

10. In technical terms, all organisms exhibit a species-specific pattern of ontogeny or individual development through the life cycle: see Stephen Jay Gould, *Ontogeny and Phylogeny* (Cambridge, Mass.: Harvard University Press, 1977). Humans are notable for the slowness of their maturation to full adult capabilities, a characteristic technically called neoteny: see Irenäeus Eibl-Eibesfeldt, *Human Ethology* (New York: Aldine de Gruyter, 1989). On the biological analysis of the life cycle and its relevance to human behavior, see Richard Alexander, *Biology of Moral Systems* (New York: Aldine de Gruyter, 1987), chap. 1. For evidence that this perspective is necessary to understand the development of the human brain, see Chive Aoki and Philip Siekevitz, "Plasticity in Brain Development," *Scientific American* 259 (1988): 56–64.

11. Thomas R. Scott and Barbara K. Giza, "Coding Channels in the Taste System of the Rat," *Science* 249 (1990): 1585–87.

12. In addition to Gazzaniga's *The Social Brain* and *Mind Matters*, see Daniel C.

Dennett, *Conscousness Explained* (Boston: Little, Braun, 1991). A generation before neuroscientists demonstrated the precise structures involved, Konrad Lorenz demonstrated the existence of "innate releasing mechanisms" as essential factors in animal social behavior. See esp. Konrad Lorenz, *Studies in Animal and Human Behaviour* (Cambridge, Mass.: Harvard University Press, 1970–71); and Eibl-Eibesfeldt, *Human Ethology*.

13. On taste receptors and the modification of their sensitivity, see Scott and Giza, "Coding Channels." Probably the best-known example of neonatal "insults" on the brain is fetal alcohol syndrome, in good part due to Michael Dorris's best-selling *The Broken Cord* (New York: Harper and Row, 1989). Less well known are similar effects of the masculine and femine sex hormones (testosterone and estrogen) on the structure and development of the brain; see Doreen Kimura and Elizabeth Hampson, "Neural and Hormonal Mechanisms Mediating Sex Differences in Cognition," Research Bulletin No. 689 (April 1990), Department of Psychology, University of Western Ontario, London, Canada.

14. See Theodosius Dobzhansky, *Mankind Evolving* (New Haven, Conn.: Yale University Press, 1955); Roe and Simpson, eds., *Evolution and Behavior*; George Gaylord Simpson, *Biology and Man* (New York: Harcourt, Brace and World, 1969).

15. At this writing, research is developing rapidly, with no firm scientific consensus on the full explanation of numerous complex traits that seem to show evidence of genetic predisposition. On Alzheimer's disease, Dennis J. Selkoe, "Amyloid Protein and Alzheimer's Disease," *Scientific American* 263 (November 1991): 68–79; on mental disorders more generally, Elliot S. Gershon and Ronald O. Rieder, "Major Disorders of Mind and Brain," *Scientific American* 267 (September 1992): pp. 127–33.

16. Jacques Mehler, "Language Comprehension: The Influence of Age, Modality, and Culture" (Paper presented at the 37th annual meeting of the Orton Dyslexia Society, Philadelphia [Inglewood, Calif: Audio-Stats Educational Services, tape no. 916R-21]).

17. Gerald Edelman, *The Remembered Present* (New York: Basic Books, 1989).

18. "These and a million of such other propositions, as many at least as we have distinct ideas of, every man in his wits, at first hearing, and knowing what the names stand for, must necessarily assent to. . . . But, since no proposition can be innate unless the *ideas* about which it is be innate, this will be to suppose that all our ideas of colors, sounds, tastes, figures, &c., innate, than which there cannot be anything more opposite to reason and experience" (*Essay*, 1.1.18; vol. 1, pp. 52–53; italics in original).

19. Robert Plomin and Denise Daniels, "Why Are Children in the Same Family So Different from One Another? *Behavioral and Brain Science* 10 (1987): 1–30; L. J. Eaves, H. J. Eysenck, and N. G. Martin, *Genes, Culture and Personality* (New York: Academic Press, 1989); Roger D. Masters, "Individual and Cultural Differences in Response to Leaders: Nonverbal Displays, Cognitive Neuroscience, and Politics," *Journal of Social Issues* 47 (1991): 151–65.

20. See Pawel Lewicki, Thomas Hill, and Maria Czyzewska, "Nonconscous Acquisition of Information," *American Psychologist* 47 (1992): 796–801, as well as Gazzaniga, *The Social Brain*. Even if, by "mind," Locke meant consciousness—and this could be questioned since he explicitly associates consciousness with the "soul," as distinct from the "mind" (e.g., *Essay*, 2.1.11; vol. 1, p. 130)—his distinc-

tion would still be inconsistent with cognitive neuroscience. In this interpretation the active properties that Locke associates only with "reflection" are essential to the entire process of processing sensory input.

21. See the evidence for the existence of an "interpreter module" described in Gazzaniga's *The Social Brain*.

22. Robert Frank, *Passions within Reason*, (New York: W. W. Norton, 1988) and "Emotion and the Costs of Altruism," in *The Sense of Justice*, ed. Roger D. Masters and Margaret Gruter (Newbury Park, Calif.: Sage, 1992), pp. 47–66.

23. For the difference between the deaf and the hearing, see Harlan Lane, *The Mask of Benevolence: Disabling the Deaf Community* (New York: Knopf, 1992). On dyslexia and learning disabilities, see Masland, "Neurological Aspects of Dyslexia," pp. 35–66.

24. Howard Gardner, *Frames of Mind* (New York: Basic Books, 1983).

25. Roger Draper, "A Voice of Silence," *The New Leader*, June 29, 1992, p. 15.

26. See, inter alia, Plato, *Republic*, esp. bks. 8 and 9; Theophrastus, *Characters*; Plutarch, *Moralia*. For a contemporary analysis that shows the importance of returning to this tradition, see James Q. Wilson, *On Character* (Washington, D.C.: AEI Press, 1991).

27. *Essay*, 1.2.10; p. 74 (cited in full in n. 8, chap. 7, above). Locke's parenthetical aside, omitted here, is of course critical for understanding the difference between the ancients and the moderns. Whereas Plato and Aristotle, not to mention Augustine or Aquinas, had no difficulty in understanding how some individuals and societies (or even most of them) might deviate from the "rules" that are "best" for human life, for Locke even a single exception seems to be sufficient to disprove the traditional derivation of morality from either nature or divine will (cf. n. 17 above). Laws of nature must be universally valid. Like the Sophists and other ancients who explain society as a contract, Locke and the moderns associate what is natural with the impulses of self-interested individuals. The difference is that, for the Sophists and other ancient conventionalists, this provided the basis of value preferences concerning the best life (the individual hedonism of Antiphon, the praise of tyranny of Thrasymachus, or the philosophic asceticism of the Epicureans), whereas the Lockean or modern perspective treats all such choices as equally valid.

28. Ibid, p. 75, cited above in chap. 2. By implication, Locke seems to require that innate knowledge entail the certainty and universality traditionally attributed only to divine omniscience in the Christian tradition.

29. Hume, *Treatise*, 3.1.2; p. 425, cited above in chap. 2, and Brecht, *Political Theory*, chap. 10, esp. p. 394. Note that Brecht asserts that such universals can be discovered empirically and even offers a list of them (p. 396), apparently ignoring the contrary evidence stressed by thinkers from Locke, Hume, or Marx to such contemporary anthropologists as Lévi-Strauss or Marvin Harris. Beneath Brecht's "scientific value relativism" one might discern an "absolute" deference to Western liberal customs and therewith a form of "cultural imperialism": cf. Lloyd Weinreb, *Natural Law and Justice* (Cambridge: Harvard University Press, 1987).

30. While Hume is also hostile to an "eternal" and "objective" moral law, he introduces an ambivalence in the use of the word *nature* (*Treatise*, 3.1.2; pp. 427–28, cited above). Hence for Hume, unlike Locke, the laws of justice are "unnatural" only in the sense of being derived from "thought or reflection" (*Treatise*, 3.2.1; p. 436, cited above); the use of thought or reflection to adopt conventional rules of justice is itself "necessary" and "natural."

31. This is particularly obvious when the Lockean tradition is contrasted with 18th-century political thought in France. See, e.g., Montesquieu, *L'esprit des lois* in *Oeuvres complètes de Montesquieu*, ed. Roger Caillois (Paris, Bibliothèque de la Pléiade, 1966), vol. 2, pp. 227–995; Rousseau, *Social Contract*, 2.8–9; 3.8 (ed. Masters; pp. 70–76, 92–95). For historicists in the tradition of Hegel and Marx, the study of human history reveals a temporal pattern that explains why norms or rules of conduct vary from one human community to another. For Lévi-Strauss and cultural anthropologists of his school, norms rest on a "deep structure" or system of fundamental symbolic oppositions that are elaborated by each human culture and embedded in myths or beliefs. Here the basic explanations arise from the articulation of such conceptual oppositions as "up/down," "inside/outside," "natural/conventional," "sacred/profane," or "kin/stranger" (and hence spatial or symbolic rather than temporal structures); e.g., Claude Lévi-Strauss, *La pensée sauvage* (Paris: Plon, 1962), *Les structures elementaires de la parenté* (Paris: Mouton, 1967). And for materialist anthropologists like Marvin Harris, as for behavioral ecologists who apply evolutionary theory to human societies as well as to nonhuman species, social norms arise from the *relationship* between a population and its environment rather than either the abstract "nature" of humans or the "environment" established by history or culture: Marvin Harris, *Cannibals and Kings* (New York: Random House, 1977); Napoleon Chagnon and William Irons, eds., *Evolutionary Biology and Human Social Behavior* (North Scituate, Mass.: Daxbury Press, 1972); Pierre van den Berghe, *Human Family Systems* (New York: Elsevier, 1979).

32. Cf. Alexander, *The Biology of Moral Systems*; Glendon Schubert, *Evolutionary Politics* (Carbondale: Southern Illinois University Press, 1989); and Robin Fox, *The Search for Society* (New Brunswick, N.J.: Rutgers University Press, 1989), as well as the references in n. 30, 62, and 68. This issue is also the focus of my forthcoming work, *The Nature of Obligation*.

33. *Essay*, 1.2.11; vol. 1, p. 75, cited above, chap. 2. Interestingly enough, abortion (along with adultery) is a traditional taboo that is shattered in Machiavelli's *Mandragola*. It might be noted in passing that the relevance of this argument might be challenged by Plato's proposals for the education and rearing of children in the best city "according to nature" in the *Republic*.

34. This example is obviously relevant to contemporary evolutionary theories according to which all organisms can be said to optimize "reproductive success" (Wilson, *Sociobiology*; Richard Alexander, *Darwinism and Human Affairs* [Seattle: University of Washington Press, 1979]; Michael Ruse, *Taking Darwin Seriously* [Oxford: Blackwell, 1986]). See "Nonverbal Behavior, Emotion, and Social Cognition," Chap. 5, above.

35. The full text of Locke's statement of the example is revealing: "Let us take any of these rules, which, being the most obvious deductions of human reason, and conformable to the natural inclination of the greatest part of men, fewest people have had the impudence to deny or inconsideration to doubt of. If any can be thought to be naturally imprinted, none, I think, can have a fairer pretence to be innate than this: 'Parents, preserve and cherish your children.' When, therefore, you say that this is an innate rule, what do you mean? Either that it is an innate principle which upon all occasions excites and directs the actions of all men; or else, that it is a truth which all men have imprinted on their minds, and which therefore they know and assent to" (*Essay*, 1.2.12; vol. 1, pp. 75–76). Locke manifestly requires,

for a rule to be "innate" or "naturally imprinted," that it either control the behavior of "all men" (even unconsciously) or that it be universally acknowledged as a binding rule. The first requisite is contradicted by a single case of child abuse; the second, by a single society that engages in generally approved infanticide. In short, Locke's method ultimately assumes that nature is determinist and invariant rather than probabilistic and variable.

36. Locke puts it as follows:

To make it capable of being assented to as true, it must be reduced to some such proposition as this: "It is the duty of parents to preserve their children." But what duty is, cannot be understood without a law; nor a law be known or supposed without a lawmaker, or without reward and punishment; so that it is impossible that this, or any other, practical principle should be innate, i.e., be imprinted on the mind as a duty, without supposing the ideas of God, of law, of obligation, of punishment, of a life after this, innate: for that punishment follows not in this life the breach of this rule, and consequently that it has not the force of a law in countries where the generally allowed practice runs counter to it, is in itself evident. (*Essay*, 1.2.12; vol. 1, p. 76).

37. On infanticide among nonhuman primates, see J. Itani, in *Law, Biology and Culture*, ed. Margaret Gruter and Paul Bohannan (Santa Barbara, Calif.: Ross Erikson, 1983), pp. 62–74; and Sara Blaffer Hrdy, "Infanticide among Animals," *Ethology and Sociobiology* 1 (1979): 13–40. For human mating systems and their relationship to evolutionary theory, see Robin Fox, ed., *Biosocial Anthropology* (London: Malaby, 1975); and Pierre van den Berghe, *Human Family Systems* (New York: Elsevier, 1979). Perhaps the clearest case is the role of female infanticide in hypergamous mating systems, a complex pattern characterizing socially stratified societies confronting wars, famines, plagues, and highly unpredictable resource flows; see Mildred Dickemann, "Female Infanticide, Reproductive Strategies, and Social Stratification: A Preliminary Model," in Chagnon and Irons, eds., *Evolutionary Biology and Human Social Behavior*, pp. 321–67.

38. See the example in Napoleon Chagnon, "Yanomamö, the True People," cited in Irenäeus Eibl-Eibesfeldt and Marie-Claude Mattel-Müller, "Yanomami Wailing Songs and the Question of Parental Attachment in Traditional Kinbased Societies," *Anthropos* 85 (1990): 508. See also Eibl-Eibesfeldt, *Human Ethology*, and Herbert Helmrich, "An Ethological Interpretation of the Sense of Justice on the Basis of German Law," in *The Sense of Justice*, ed. Roger D. Masters and Margaret Gruter, Newbury Park, Calif.: Sage, 1992), chap. 10.

39. James Q. Wilson, "The Moral Sense" (Presidential address to 1992 Annual Meeting of the American Political Science Association, Chicago, Sept. 3–6, 1992). For a recent theoretical interpretation of parental bonding and infant attachment, see G. W. Kraemer, "A Psychobiological Theory of Attachment," *Behavioral and Brain Research* 15 (September 1992): 493–541.

40. Behavioral ecologists call the first of these patterns an *r*-strategy, whereas the second is termed a *K*-strategy: the difference has sometimes been conveniently illustrated by fruitflies (many offspring, little parental investment in each) and elephants (few offspring, much parental investment in each). For the principles involved, see Wilson, *Sociobiology*, David Barash, *Sociobiology and Behavior*, 2nd ed. (New York: Elsevier, 1982), or Robert Trivers, *Social Evolution* (Menlo Park, Calif.: Benjamin/Cummings, 1985). For the application to humans, see also Dickemann,

"Female Infanticide, Reproductive Strategies, and Social Stratification" and Masters, "Explaining "Male Chauvinism" and "Feminism": Cultural Differences in Male and Female Reproductive Strategies," in *Biopolitics and Gender*, ed. Meredith Watts (New York: Haworth Press, 1984), pp. 165–210.

41. In addition to the citations above, see T. D. Weldon, *The Vocabulary of Politics* (Harmondsworth, England: Penguin, 1953), esp. chap. 2: "In the case of political theories the title 'classical' is especially appropriate because the premises in question were here inherited, *unchanged in all essentials*, from the Greek writers of the classical period. They are not at all satisfactory, and it is hard to believe that, if the Greek civilization had endured with anything like its initial vitality, they would not have been criticized and discarded" (p. 17, italics added). Weldon refers to theorists from Plato to Machiavelli, Rousseau, and Marx.

42. Aristotle describes such mutuations or accidents as frequent events in which "nature fails to achieve her intention"; this usage has confused modern readers who do not understand his conception of teleology. See Larry Arnhart, "Aristotle's Biopolitics: A Defense of Biological Teleology against Biological Nihilism," *Politics and the Life Sciences* 6 (1988): 173–229; and "Aristotle, Chimpanzees, and Other Social Animals," *Social Science Information* 19 (1990): 479–559. It is interesting to contrast Hume's treatment of child abuse with that of Locke, since for Hume the duty to care for children is evidence of the need to distinguish between natural impulses or inclinations and the duties that can arise from them: "We blame a father for neglecting his child. Why? because it shows a want of natural affection, which is the duty of every parent. Were not natural affection a duty, the care of children could not be a duty; and it were impossible we could have the duty in our eye in the attention we give to our offspring. In this case, therefore, all men suppose a motive to the action distinct from a sense of duty" (*Treatise*, 3.2.1; p. 431). On the basis of this argument, Hume can establish "natural virtues" (like benevolence) as well as "artificial" or "cultural" ones (like justice).

43. James B. Murphy, "Nature, Custom, and Stipulation in Law and Jurisprudence," *Review of Metaphysics* 43 (1990): 751–90. Cf. Masters, *The Nature of Politics*, chap. 4, and Arlene Saxonhouse, *Fear of Diversity: The Birth of Political Science in Ancient Greek Thought* (Chicago: University of Chicago Press, 1992).

44. Brecht's discussion of the Darwinian theory of evolution is telling:
Scientific Method must . . . insist that when we speak of "goal" or "purpose" and of their embodiment or the like, we must say what we mean by these terms. It must point out that, if we mean by them, *as we generally do*, ideas which some mind has formed and pursues, then the assumption of goals or purposes in plants, seeds, sperms, eggs, organs, etc. *implies the assumption of a superhuman mind.* . . . Scientific Method must insist, on the other hand, that the reality of natural evolution is not proved by a demonstration of its theoretical possibility alone. Even if fully verified, natural evolution would fail to explain how the entire chain of events that led to such miraculous results was set going and to establish whether or not that was done by a Creator with a purpose—a purpose incorporated, not perhaps in individual things, but in the entire chain of events and the evolutionary laws governing them. (Brecht, *Political Theory*, pp. 87–88; italics added)
As this passage makes clear, Brecht views teleological reasoning as somehow tantamount to postulating a "superhuman mind" or "Creator with a purpose"—i.e., he

does not take into consideration Aristotle's view of the nature of a thing as its end or purpose in a cosmos without the Creative God of Genesis. In considering this passage, it is also worth reflecting on Brecht's poor understanding of evolutionary biology.

45. This is particularly the case because, with the advent of birth control techologies controlled by women, abortion has taken on a new function in assuring that women can control their access to the labor market. See Lionel Tiger, *The Manufacture of Evil* (New York: Harper & Row, 1987), chap. 6.

46. Note that historical, social, and technological factors—rather than free choice—establish what is "the better among those that are possible" (Plato, *Republic*, 10.618c). Because both the right to life and the right to choose take one of the stages in this sequence of preferences and make it an absolute, seeking a universal rule on the model of Kantian ethics, modern formulations deny the role of prudence and generate intractible conflict.

47. Gina Kolata, "Genetic Defects Detected in Embryos Just Days Old," *New York Times*, September 24, 1992, pp. A1, B10. In cystic fibrosis, a person is physically affected only if the gene has been inherited from both parents. In this case, each of the parents carried the gene on only one chromosome; when both parents have a single copy of such a gene, each of their children has one chance in four of inheriting it from both parents and hence of being a victim of the genetic disease.

48. It should be obvious—but is rarely noted—that this mode of thought coincided with the emergence of the market economy and modern states governed by bureacracies according to what Weber called "rational-legal legitimacy." In a government under law, the legislature enacts legal rules, and bureaucrats are expected to administer them without favoritism ("If I did that for you, I would have to do it for everyone"). From the perspective of the history and philosophy of science, the conceptualization of "laws of nature" thus does indeed mirror (and subtly justify) developments in the political and legal sphere: compare chaps. 2 and "Intuition and the Critique of Science" in chap. 4 with the citations in n. 6, chap. 2.

49. This pattern of thought helps explain the otherwise puzzling similarity of neo-Kantian moralists, analytic philosophers, and logical positivists—all of whom reject the attempt to identify the natural basis of moral principles. It is ironic to note that this emphasis on abstract rules as the highest form of moral reasoning has been attributed to patterns of thought more likely to be exhibited by males than by females: see Carol Gilligan, *In a Different Voice* (Cambridge: Harvard University Press, 1983). Some neuroscientists suggest that the gender differences emphasized by Gilligan may derive from differences in the structure and function of the brain due to hormonal effects during development; e.g., Kimura, "Sex Differences in the Brain." The possibility that Kantian metaphysics and ethics might reflect patterns of thought that are more typical of males than of females indicates clearly how difficult it is to claim that ethical principles can be established without reference to scientific facts.

50. Mortimer Mishkin and Thomas Appenzeller, "The Anatomy of Memory," *Scientific American* 256 (1987). It is true that established memory is associated with lasting changes in the cortex such that emotional activation may no longer be involved in some forms of memory; see Stuart M. Zola-Morgan and Larry R. Squire, "The Primate Hippocampal Formation: Evidence for a Time-Limited Role in Memory Storage," *Science* 250 (1990): 288–90. But this would explain estab-

lished habit, not the rational judgments or formation of habits that are central for Kant.

51. Once "natural kinds" are admitted into the argument, there is no obvious reason to reject naturalism. For the critique of neo-Kantian thought on these grounds, see (among others) Ruse, *Taking Darwin Seriously*; Shapiro, *Political Criticism*; and Dennett, *Consciousness Explained*. For Kant himself, "independence" or "autonomy" of the will were essential criteria, allowing the exclusion of the indentured laborer (or women) from the class of moral agents. But as the fictional computer HAL illustrated, it is clearly possible to imagine a machine engaging in an independent act (and, indeed, many computer users already wonder, only partly in jest, whether their machines malfunction "on purpose"). The key question, of course, concerns conscious intentionality, and consciousness in this sense is usually defined in terms of human experience. The same problem arises, incidentally, with regard to the possibility that nonhuman primates meet Kantian criteria of intentionality or autonomy in nonhuman primates (Daniel C. Dennett, "Précis of *The Intentional Stance*" and peer commentary, *Behavioral and Brain Sciences* 11 [1988]: 495–546; Roger D. Masters, "Conclusion," in *Primate Politics*, ed. Glendon Schubert and Roger D. Masters [Carbondale: Southern Illinois University Press, 1991], esp. pp. 235–39).

52. Roger Penrose, *The Emperor's New Mind* (New York: Oxford University Press, 1989).

53. Ibid., chap. 1. The relationship between the logic of the "stop" command and the discussion of pattern matching in chap. 4 is of the greatest importance.

54. See Stuart Zola-Morgan, Larry Squire, Pablo Alvarez-Royo, and Robert Clower, "Independence of Memory Functions and Emotional Behavior: Separate Contributions of the Hippocampal Formation and the Amygdala," *Hippocampus* 1 (1991): 207–20.

55. Penrose, *The Emperor's New Mind*, pp. 102, 104. For a restatement of Gödel's theorem, see pp. 105 ff.

56. Arnold Brecht, *Political Theory*, p. 56 and n. 2.

57. Brecht should have known this. Consider Penrose's account:
Russell himself, together with his colleague Alfred North Whitehead, set about developing a highly formalized mathematical system of axioms and rules of procedure, the aim being that it should be possible to translate all types of correct mathematical reasoning into their scheme. . . . The specific scheme that Russell and Whitehead produced was a monumental piece of work. However, it was very cumbersome, and it turned out to be rather limited in the types of mathematical reasoning that it actually incorporated. The great mathematician David Hilbert . . . embarked on a much more workable and comprehensive scheme . . . the hopes of Hilbert and his followers were dashed when, in 1931, . . . Gödel produced a startling theorem. (Penrose, *The Emperor's New Mind*, pp. 101–102).
In short, when Brecht wrote in 1958, he cited the work of Russell and Whitehead (published in 1919) rather than that of Hilbert, which itself was fundamentally challenged—and in the eyes of mathematicians demolished—by Gödel's publication in 1931. It is hard to avoid the impression that Brecht's knowledge of logic was as derivative as is my own! If nothing else, this may suggest the practical difficulties of basing the methodology of social science on highly technical logical theories that

are incomprehensible to all but specialists and the mathematically gifted; for most of those who read Brecht's *Political Theory*, these issues were not "intersubjectively verifiable."

58. Cf. Roger D. Masters, "Can Physics Explain Human Consciousness?" *Quarterly Review of Biology* 65 (December 1990): 481–84.

59. Leo Strauss, Introduction, *Natural Right and History* (Chicago: University of Chicago Press, 1953), pp. 7–8.

9. Integrating Nature and Nurture

1. Carl Degler, *In Search of Human Nature* (New York: Oxford University Press, 1991).

2. See Roy Bhaskar, *The Possibility of Naturalism* (Atlantic Highlands, N.J.: Humanities Press, 1979), and *Scientific Realism and Human Emanicipation* (London: Verso Books, 1986).

3. James Gleick, *Chaos* (New York: Viking, 1987); David Ruelle, *Chance and Chaos* (Princeton, N.J.: Princeton University Press, 1991); and, for more details, chap. 3 above.

4. Cf Robert M. May, "How Many Species?" *Philosophical Transactions of the Royal Society of London*, ser. B, 330 (1990): 293–304. As May shows, the ecological niches of many species are not limited by our three-dimensional (or two-dimensional) conceptions of the world. To understand what this means, try to imagine a geometric shape in 2.6 dimensions!

5. A social scientist who speaks of a "nonlinear" relationship between two variables refers to a smooth curve, in which the value of one variable first increases and then decreases as a factor of the other; the term is shorthand for the statistical technique of nonlinear regression. In statistics, a regression equation is an inductive technique that can be visualized as the line through a set of points minimizing the error of predicting each datum of the dependent variable on the basis of the specified independent variables. Political scientists and sociologists have typically employed linear regression, seeking to fit a straight line through the data points being analyzed; analysis of variance provides a mathematically similar approach more widely used in psychology. The assumptions underlying these methods illustrate the way nominalism (reflected in the use of inductive statistics) and Newtonian physics (reflected in the presumption of inertia and linear causation) are deeply, though usually unconsciously, embedded in the conventions of empirical social scientists. E.g., Leona Aiken and Stephen G. West, *Multiple Regression: Testing and Interpreting Interactions* (Newbury Park, Calif.: Sage, 1991); Douglas C. Montgomery and Elizabeth Peck, *Introduction to Linear Regression Analysis*, 2nd ed. (New York: Wiley, 1992).

6. For characteristic texts for courses in the "scope and methods of political science," see the works cited in n. 2, chap. 2 above. According to the widely cited model of Hempel and Oppenheim, the scientific explanation of an event entails the ability to subsume it under a "general" or "covering law," normally stated in the hypothetical form: "If x, then y." In principle, prediction and postdiction based on scientific hypotheses should be reciprocal—i.e., one should be able to use the same methods to explain the past and to predict the future (Carl G. Hempel and Paul Oppenheim. "The Logic of Explanation," *Philosophy of Science* 15 [1948];

Carl Hempel, "The Function of General Laws in History," *Journal of Philosophy* 39 [1942]: 35–48). At least in the Popperian view, the test of hypotheses is their falsifiability if not their actual falsification (Karl Popper, *The Logic of Scientific Discovery* [London: Hutchinson, 1959]). I am, of course, aware of the radical critiques of this approach by some philosophers of science (e.g., Paul Feyerabend, *Against Method* [London: NLB; Altantic Highlands, N.J.: Humanities Press, 1975]). Because empirically oriented social scientists (as distinct from postmodern humanists) usually pay little or no attention to these critics, their arguments can hardly be used as evidence concerning the philosophy of science underlying mainstream sociology, political science, economics, or psychology.

7. Theodore L. Becker, ed., *Quantum Politics: Applying Quantum Theory to Political Phenomena* (New York: Praeger, 1991).

8. Stephen Hawking, *A Brief History of Time* (New York: Bantam, 1988); Werner Heisenberg, *Physics and Philosophy* (New York: Harper Torchbooks), 1958); Roger Penrose, *The Emperor's New Mind* (New York: Oxford University Press, 1989). This is, it needs to be added, not the only interpretation possible. Other physicists, following Bohr's interpretation of uncertainty, have taken views more closely related to the postmodernist or deconstructionist movement in the humanities (e.g., Berlinski, *Black Mischief: Language, Life, Logic, Luck*, 2nd ed. [Boston: Harcourt Brace Jovanovich, 1988]). Such views should be of little comfort to traditional social scientists, however, since they lead to an extreme relativism and a denial of all forms of objective or scientific knowledge (see chap. 4, above).

9. For an application of quantum mechanics to the human brain and thought, however, see Henry Pierce Stapp, "A Quantum Theory of the Mind-Brain Interface," Lawrence Berkeley Laboratory Report LBL–28574 Expanded (1990).

10. For the concept of "reaction range" or "norm of reaction" in varied environments, see Theodosius Dobzhansky, *Mankind Evolving* (New Haven; Conn.: Yale University Press, 1955), George Gaylord Simpson, *The Meaning of Evolution*, rev. ed. (New Haven, Conn.: Yale University Press, 1967). On the reorganization of the human brain during development, see Chive Aoki and Philip Siekevitz, "Plasticity in Brain Development," *Scientific American* 259 (1988): 56–64. For illustrations, see Roger D. Masters, *The Nature of Politics* (New Haven, Conn.: Yale University Press, 1989), chap. 4.

11. For the study of thoroughbred horses, which found the heritability of racing speed to be .35, see Patrick Cunningham, "The Genetics of Thoroughbred Horses," *Scientific American* 264 (1991: 92–98. This example is particularly telling for three reasons: First, the observed performance has been exactly measured (if only because of its importance for those who bet on races); second, the hereditary lineages have been carefully recorded (because the monetary value of a thoroughbred depends in part on it); and finally, the data reflect examination of the *entire universe* of thoroughbred horses rather than merely a sample (thereby excluding the complaint that sampling error explains the role of heritability in complex social traits). For human personality, see L. J. Eaves, H. J. Eysenck, and N. G. Martin, *Genes, Culture, and Personality: An Empirical Approach* (New York: Academic Press, 1989) and Robert Plomin, "The Role of Inheritance in Behavior," *Science* 248 (1990): 183–88) as well as MacDonald, *Social and Personality Development: An Evolutionary Synthesis* (New York: Plenum, 1989). Since the dimensions of personality distinguished in these studies have also been observed in vervet monkeys (Michael T. McGuire, unpublished data) and there are plausible mechanisms by which personality traits

could be transmitted genetically yet modified by experience (see n. 24 below), there is no reason to apply to these findings the strictures usually raised against the analysis of heritability in IQ (a trait that, it can be argued, does not measure intelligence in any socially meaningful sense). For a clear explanation of the concept of heritability and a defense of its application to IQ testing, see Richard Herrnstein, "IQ," *Atlantic Monthly*, September 1971, pp. 43–64; for the critique of these heritability studies with particular focus on IQ scores, see Leon J. Kamin, *The Science and Politics of I.Q.* (Potomac, Md.: Lawrence Erlbaum, 1974); and Richard Lewontin, Steven Rose, and Leon Kamin. *Not in Our Genes* (Boston: Pantheon, 1984). With regard to nonverbal intelligence (one skill that is most likely to be reasonably measured by the highly controversial IQ test), however, statistical correlations between the test scores of the *offspring* of identical twins and their twin aunt/uncle or that twin aunt/uncle's spouse show results similar to those in studies of traits whose inheritence is unquestioned; see Richard J. Rose, E. L. Harris, J. C. Christian, and W. E. Nance, "Genetic Variance in Nonverbal Intelligence: Data from the Kinships of Identical Twins," *Science* 205 (1979): 1153–54. This study is particularly interesting because it includes, as a control, the study of fingerprint whorls—one of the relatively few human phenotypical traits that seem to be entirely controlled by the genotype.

12. This point, which follows from the elementary principles of statistics, means that it is nonsense to assume that population studies of heritability are demonstrations of "genetic determinism." Public misperception of this field, encouraged by the erroneous nature–nurture dichotomy, leads to a facile equation of the concepts of "biological," "genetic," and "heritability." Cf. Richard Alexander, "Biology and Law," in *Ostracism*, ed. Margaret Gruter and Roger D. Masters, (New York: Elsevier, 1986) pp. 19–25. Ultimately, the ancient distinction between the one and the many (see above, chap. 3 "The One and the Many") is critical when applying a scientific approach to individual cases.

13. On chimpanzees: Jane Goodall, *The Chimpanzees of Gombe: Patterns of Behavior* (Cambridge, Mass.: Harvard University Press, 1986). There is increasing evidence that sex-linked hormones influence neuronal structures during fetal development and infancy, producing some of the differences between human males and females that have been so widely noted. For the neurochemistry of development, see Lee Ellis, "Evidence of Neuroandrogenic Etiology of Sex Roles from a Combined Analysis of Human, Nonhuman Primate and Nonprimate Mammalian Studies," *Personality and Individual Differences* 7 (1986): 519–52 and, specifically on humans, John Money and Anke Ehrhardt, *Man and Woman, Boy and Girl: The Differentiation and Dimorphism of Gender Identity from Conception to Maturity* (Baltimore: Johns Hopkins University Press, 1972). While this field is controversial, it is hardly fair to assume—as some have done—that the observation of gender differences necessarily implies that women are inferior; on the contrary, the notion that differences in social behavior between males and females have a natural as well as a cultural component might point to the superiority of females on many politically relevant traits. See Glendon Schubert, *Sexual Politics and Political Feminism* (Greenwich, Conn.: JAI Press, 1991); and for an illustration of recent work in social psychology, Philip Shaver and Clyde Hendrick, eds., *Sex and Gender: Review of Personality and Social Psychology* (Beverly Hills, Calif.: Sage, 1987), vol. 7. The refusal of some scholars to entertain this possibility gives the impression that

ideological considerations are involved. Compare "Facial Displays and Political Leadership" in chap. 5, above.

14. Given the enormous literature that has followed upon the publication of Edward O. Wilson's *Sociobiology* (Cambridge, Mass.: Harvard University Press, 1975) and Richard Dawkins' *The Selfish Gene* (New York; Oxford University Press, 1976), there is little need to say more about the fundamental principles. For those who find technical scientific writing aversive, an easily accessible introduction is Christopher Badcock, *Evolution and Individual Behavior: An Introduction to Human Sociobiology* (Oxford: Blackwell, 1991). What needs emphasis, however, is that up-to-date studies show the way both ecological and social conditions shape the strategies of organisms; for an excellent review, see Robert L. Trivers, *Social Evolution* (Menlo Park, Calif.: Benjamin/Cummings, 1985). Those who apply findings from this field to humans need to be aware that the concept of "behavioral ecology" fits its methods and findings more accurately than does "sociobiology" (a word whose connotations of genetic determinism represent ideological misperceptions rather than scientific research). See "The Origins of Society" in chap. 6 above and "Cultural Relativism: Morality, Virtue, and Obligation" in chap. 8 above.

15. E.g., Jack Hirshleifer, "Economics from a Biological Viewpoint," *Journal of Law and Economics* 20 (1977): 1–52, and *Economic Behavior in Adversity* (Chicago: University of Chicago Press, 1987); Robert Frank, *Passions within Reason* (New York: W. W. Norton, 1988); Gordon Getty, "The Hunt for *r*: One-Factor and Transfer Theories," *Social Science Information* 28 (1989): 385–428. This confluence of economic and evolutionary perspectives has been fruitfully applied to the development of legal and ethical concepts; see Paul Rubin, "Evolved Ethics and Efficient Ethics, *Journal of Economic Behavior and Organization* 3 (1982): 161–74; E. Donald Elliott, "Holmes and Evolution: Legal Process as Artificial Intelligence," *Journal of Legal Studies* 13 (1984): 13; Robert Cooter, "Inventing Market Property: The Land Courts of Papua New Guinea," *Law and Society Review* 25 (1991): 759–95.

16. Robert Frank, *Choosing the Right Pond* (New York: Oxford University Press, 1985). In general, the theories of economists, psychologists, sociologists, and political scientists measure the absolute value of things: the cost of goods or net benefit of resources, power, and status (rather than the value *relative* to either other members of the group or to the individual's own expectations). One of the rare exceptions not derived from evolutionary theory is the notion of framing and risk assessment, introduced by Kahneman and Tversky and now known as Prospect Theory. See Daniel Kahneman and A. Tversky, "Prospect Theory: An Analysis of Decision under Risk," *Econometrica* 47 (1979): 263–91; and for a review of the theory's implications, Barbara Farnham, ed., "Prospect Theory and Political Psychology," *Political Psychology* (Special Issue) 13 (1992): 167–329.

17. Mildred Dickemann, "The Ecology of Mating Systems in Hypergynous Dowry Societies, *Social Science Information* 18 (1979): 163–95; Napoleon Chagnon and William Irons, eds., *Evolutionary Biology and Human Social Behavior* (North Scituate, Mass.: Duxbury Press, 1979); Richard A. Posner, *Sex and Reason* (Cambridge: Harvard University Press, 1992); Masters, "Explaining 'Male Chauvinism' and 'Feminism': Cultural Differences in Male and Female Reproductive Strategies," in *Biopolitics and Gender*, ed. Meredith Watts (New York: Haworth Press, 1984), pp. 165–210. See also n. 14 above.

18. For models of cultural evolution, see Charles J. Lumsden and Edward O. Wilson, *Genes, Mind, and Culture* (Cambridge, Mass.: Harvard University Press, 1981); Robert Boyd and Peter J. Richerson, *Culture and the Evolutionary Process* (Chicago: University of Chicago Press, 1985); and William Durham, *Coevolution: Genes, Culture, and Human Diversity* (Stanford, Calif.: Stanford University Press, 1991). On the concept of "evolutionary epistemology," see James H. Fetzer, ed., *Sociobiology and Epistemology* (Dordrecht: Reidel, 1985).

19. Ernst Mayr, *Species and Evolution* (Cambridge, Mass.: Harvard University Press, 1986). As Konrad Lorenz showed, one can distinguish between species by their behavior as well as by their physical shape: *Studies in Animal and Human Behaviour* (Cambridge, Mass.: Harvard University Press, 1970–71). This means that animals themselves treat each species as a "natural kind." How can the difference between a dog and a cat be entirely in the mind of the human beholder if both dogs and cats behave on the basis of the same discriminations embedded in most human languages?

20. The basic distinction between proximate and ultimate causation found in elementary textbooks in evolutionary biology (see the references in nn. 10 and 14, this chapter) reflects this difference between the species as a naturally occurring unit of organization and the organism as a unit of selection. These distinctions cannot be dismissed as an artifact of the scientist's perception, as many epistemologists have claimed; different methods and phenomena are needed to test hypotheses concerning proximate causes or mechanisms (responses of the organism that have been selected to optimize individual reproductive success) and ultimate causes (the system of functions or purposes that can be analyzed at the species, phylogenetic, or ecosystemic levels).

21. The theoretical conception was formulated by Gould and Eldredge, "Punctuated Equilibria: The Tempo and Mode of Evoluton Reconsidered," *Paleobiology* 3 (1977): 115–51; for a review, see Albert Somit and Steven A. Peterson, eds., *The Dynamics of Evolution: The Punctuated Equilibrium Debate in the Natural and Social Sciences* (Ithaca, N.Y.: Cornell University Press, 1992).

22. Luigi Luca Cavalli-Sforza, "Genes, Peoples and Languages," *Scientific American* 265 (1992): 104–10. For the formal parallels between genetic and linguistic systems as carriers of information, see Dawkins, *The Selfish Gene* and Masters, *The Nature of Politics*, chap. 3.

23. Michael Gazzaniga, *The Social Brain* (New York: Basic Books, 1985), and *Mind Matters* (Boston: Houghton Mifflin, 1988); Paul MacLean, "A Triangular Brief on the Evolution of Brain and Law," in *Law, Biology and Culture*, 2nd ed., ed. Margaret Gruter and Paul Bohannan (New York: Primis/McGraw-Hill, 1992), pp. 83–97; Mortimer Mishkin and Thomas Appenzeller, "The Anatomy of Memory," *Scientific American* 256 (1987). Compare "Emotion, Associative Learning, and Memory" in chap. 5, above.

24. Roger D. Masters and Michael T. McGuire, eds., *The Neurotransmitter Revolution* (Carbondale: Southern Illinois University Press, 1993). This book, arising from a Gruter Institute conference, "Serotonin, Social Behavior and the Law," which took place at Dartmouth College in 1988, surveys the effects of serotonin and other neurotransmitters in such diverse human behaviors as seasonal affective disorder, suicide, and homicide and arson, as well as in dominance behavior. The relationships between genetic predispositions, individual life histories, and social contexts are complex, demolishing the simplistic nature–nurture dichotomy. But

they also have very unsettling implications for our system of laws and moral responsibility. Consideration of an important issue such as the proposal that crime be viewed as a medical condition can hardly be constrained by obsolete methodological considerations of "levels of analysis."

25. In addition to the works cited in n. 23, above, see William Allman, *Apprentices of Wonder* (New York: Bantam, 1989); Daniel C. Dennett, *Consciousness Explained* (Boston: Little, Brown, 1991); Stephen M. Kosslyn and Olivier Koenig, *Wet Mind* (New York: Free Press, 1992); and George Johnson, *In the Palaces of Memory* (New York: Vintage, 1992).

26. Edmund T. Rolls, "The Processing of Face Information in the Primate Temporal Lobe," in *Processing Images of the Face*, ed. V. Bruce and M. Burton (London: Ablex, 1989). For an examination of the practical consequences of these findings, see "Cognitive Neuroscience and the Modular Brain" and "Emotion, Associative Learning, and Memory" in chap. 5, above.

27. These may seem strong words. Many scholars using new approaches receive anonymous reviews in which the critic's interpreter module has been pushed to rather elaborate extremes to justify rejection of articles that have been subsequently published and widely cited in the media as well as in scientific journals. This is, of course, the standard experience for those involved in research during the process of what Kuhn calls a paradigm shift (see *The Structure of Scientific Revolutions*). For examples in the history of physics, see Ruelle, *Chance and Chaos*, pp. 86–87, 103, 109–10 and Miller, *Fact and Method*, esp. pp. 506–14 ("The Dogmatism of Skepticism").

28. It is also based on theoretical grounds, which have been extremely well stated by Ian Shapiro (in *Political Criticism*) as well as by Bhaskar (see n. 2, this chapter). The relationship between these philosophical questions and ethical or political theory is the focus of my Conclusion.

Conclusion

1. While this consequence flows directly from the nonlinearity emphasized by chaos theory, it also is implicit in the historical and moral limitations of modern society that arise from the human response to science. To show this more fully, however, would require an extensive discussion of the problem of reflexivity, which the moderns can solve in principle but not effectively in practice. Compare Herbert A. Simon, "Bandwagon and Underdog Effects of Election Predictions," *Public Opinion Quarterly* 18 (1954): 245–53, with Roger Buck, "Reflexive Predictions," *Philosophy of Science* 30 (1963): 359–69. On the prediction and control of natural necessity as a characteristic of modern thought generally, see Roger D. Masters, "Nature, Human Nature, and Political Thought," in *Human Nature in Politics*, ed. Roland Pennock and John Chapman (New York: New York University Press, 1977), pp. 69–110.

2. For a careful account of this problem in the context of governmental policies in the United States, see Bruce L. R. Smith, *The Advisors: Scientists in the Policy Process* (Washington, D.C.: The Brookings Institution, 1992). This does not mean scientists are poor policymakers: on these issues, C. P. Snow's classic, *Science and Government* (Cambridge, Mass.: Harvard University Press, 1961) remains highly valuable.

3. This controversy is evident in such diverse issues as the "war on drugs," pornography, patriotism (burning the flag), the traditional family, prayer in the public schools, sexual equality, and affirmative action. While the controversy over abortion illustrates the intensity of these moral issues in society at large, their political relevance was symbolized during the early phases of the 1992 American Presidential campaign when the Republican vice president Dan Quayle openly attacked the "cultural" and "academic" elite in criticizing the "Murphy Brown" television program, while the Democratic candidate, Bill Clinton, explicitly defended the "old values" of family and community in his acceptance speech.

4. Although the most trenchant critique is probably Allan Bloom's *Closing of the American Mind* (New York: Simon and Schuster, 1986), the issues in the philosophy of science (see chap. 4 and 6) underlie the conflicts within our universities to which most attention has been directed.

5. See "Technology, Desire, and Society," chap. 2, above. As behavioral ecology shows, animals are less likely to cooperate when living in a secure and stable environment that provides abundant, readily available, and highly predictable resources. See Eric A. Smith and Bruce Winterhalder, *Evolutionary Ecology and Human Behavior* (New York: Aldine de Gruyter, 1992) and the references in nn. 14 and 17 to ch. 9. Machiavelli describes clearly the mechanisms of the resulting cycle in human history: "For virtue gives birth to quiet, quiet to leisure, leisure to disorder, disorder to ruin" (*Florentine Histories*, 5.1, trans. Laura Banfield and Harvey Mansfield, Jr. (Princeton, N.J.: Princeton University Press, 1988), p. 185.

6. Masters, "Nature, Human Nature, and Political Thought." For two exceptions to the tendency to speak of "rules of nature" rather than "laws of nature," see Plato, *Gorgias*, 83e, and *Timaeus*, 483e. On the emergence of the modern conception of "laws of nature," see n. 6, to chap. 2, above.

7. For Aristotle, nature concerns things that *either* "always come to pass in the same way" *or* "for the most part" (*Physics*, 2.5.196b); "mistakes"—in the sense of outcomes contrary to the rule—"are possible in the operations of nature" (*Physics*, 2.8.199a). Regularity within cultural or human sphere is similar, therefore, to natural phenomena: "Since in the realm of nature occurrences take place which are even contrary to nature, or fortuitous, the same happens *a fortiori* in the sphere swayed by custom" (*De Memoria et Reminiscentia*, 2.452b).

8. In contemporary biology, study of the "function" of a bodily structure, behavior, or entire organism is indispensable. Although critics of teleology typically equate function or purpose with belief in a divine plan, naturalism and revealed religion are as different as reason and faith: to avoid confusing these two, some scientists and philosophers now speak of teleonomy or purposive systems in analyzing evolutionary adaptations. (Mario von Cranach, ed., *Methods of Inference from Animal to Human Behavior* [The Hague: Mouton, 1976]). Cf. Aristotle, *Nicomachean Ethics*, 10.5.1176a. As it relates to humans, therefore, the ancient view of nature as an end or standard cannot be dismissed as self-confirming mysticism or naïve confusion.

9. Although most histories of political philosophy stress the differences between Plato and Aristotle, in the broad panorama of ancient science and philosophy their resemblances are far greater than their differences. For ancient evidence of this, see Cicero's account of Greek thought, written before the editing of the works of Aristotle by Andronicus of Rhodes, is especially valuable: see Roger D. Masters,

"The Case of Aristotle's Missing Dialogues," *Political Theory* 5 (1977): 31–60, and "On Chroust: A Reply," *Political Theory*, 7 (1979): 545–47.

10. Aristotle, *Physics*, 199b. On Aristotle's biology, see Arnhart, "Aristotle, Chimpanzees, and Other Social Animals," as well as Roger D. Masters, "Classical Political Philosophy and Contemporary Biology," in *Politikos*, Vol. 1, ed. Kent Moors (Pittsburgh: Duguesne University Press, 1989), pp. 27–40, and above, chap. 6.

11. Until recently, however, commentators have generally misunderstood Aristotelian biology because it has been assumed that he was concerned with questions of phylogeny (the origin and evolution of species), when in fact he was focused on ontogeny (the transmission and development of species-typical traits). When reread with care, the passages in Aristotle's works that I have cited not only come close to an understanding of the principle of heredity in a sense compatible with that of modern science but also show an awareness of ecological and ethological variations that were largely ignored by Hobbes, Descartes, Locke, Hume, and other modern philosophers. See Leon Kass, *Toward a More Natural Science: Biology and Human Affairs* (New York: Free Press, 1985).

12. Roger D. Masters, "The Duties of Humanity: Legal and Moral Obligation in Rousseau's Thought," in *Constitutional Democracy: Essays in Comparative Politics*, ed. Fred Eidlin (Boulder, Colo: Westview, 1983), pp. 83–105, *The Nature of Politics* chap. 7 and Epilogue, and "Evolutionary Biology and Political Theory," *American Political Science Review* 84 (1990): 195–210.

13. Diverse and conflicting perceptions, judgments, and political opinions, which moderns following Hobbes and Locke took as evidence that humans are naturally asocial or selfish, are an essential attribute of our species. The mixture of competition and cooperation, characteristic of other social animals, is for humans the focus of dialogue and possible agreement as well as deception and disagreement. That many individuals are neither just nor moderate nor wise confirms (rather than contradicts) the ancient view that these virtues are the natural end or purpose of human thought and social activity (Leo Strauss, "Classical Political Philosophy," in Strauss, *What Is Political Philosophy?* (Glencoe, Ill.: Free Press, 1959), pp. 78–94. For a more recent statement in a somewhat different theoretical tradition, see Lloyd Weinreb, *Natural Law and Justice* (Cambridge, Mass.: Harvard University Press, 1987).

14. *Politics*, 4.1.1288b10–20; 7.2.1324b25–35. See also Robert Augros and George Stanciu, *The New Biology: Discovering the Wisdom in Nature* (Boston: Shambala, 1987) and Kass, *Toward a More Natural Science*.

15. The patient's desire to be healthy is a "fact" that is known (or knowable) by the doctor. But if so, the philosopher or scientist can also seek to understand why this fact came into existence. In so doing, what can be explained quite reasonably as a consequence of natural selection also indicates why some humans, whether suffering from clinical depression or from painful disease, might actively wish to die. In contrast, the "natural right" to self-preservation of the tradition of Hobbes, Locke, and Rousseau does not explain why the willingness to die might arise as anything other than madness or a defect due to passion.

16. Hypochondria provides an apparent exception that seems to reflect the secondary benefits of sympathy from others and the fear that death is even worse than pain; as Woody Allen's films remind us, such behavior appears comic rather than

tragic because normal observers imagine that, in the place of the hypochondriac, they would "know better" than to prefer illness to health.

17. For example, there is now good evidence linking abnormally low levels of serotenergic activity in the neurochemistry of suicidal individuals: Dan Stein and Michael Stanley, "Serotonin and Suicide," in Roger D. Masters and Michael T. McGuire, eds., *The Neurotransmitter Revolution* (Carbondale: Southern Illinois University Press, 1993). On the defects of a contractual view of medical ethics, see Roger D. Masters, "Is Contract an Adequate Basis for Medical Ethics," *Hastings Center Report* 5 (1975): 24–28.

18. Pascal, *Pensées* (esp. 5, 7, 11, and 512), ed. Louis Lafuma (Paris: Editions du Seuil, 1962), pp. 35–37, 244–47. Pascal's French is idiomatic, making translation particularly difficult; the term *esprit de finesse* could also be rendered "intuitive mind," "precise thinking," or perhaps even "subtle thought." By "mathematical thinking," Pascal means what I have earlier called modern science; in this mode of thought, "the principles are empirically perceptible but far from common usage, so that it is difficult to turn one's mind in that direction because of the lack of habit; but as soon as one turns it that way, one sees its principles fully, and it would be necessary to have completely false thinking to reason badly on principles so obvious that it is almost impossible to miss them (*[dans l'esprit géométrique] "les principes sont palpables mais éloignés de l'usage commun de sorte qu'on a peine à tourner la tête de ce côté-la, manque d'habitude: mais pour peu qu'on l'y tourne, on voit les principes à plein; et il faudrait avoir tout à fait l'esprit faux pour mal raisonner sur des principes si gros qu'il est presque impossible qu'ils échappent"* [Pensée 512; p. 244]). As thinkers from Bacon to Ortega have noted, modern scientific methods permit even ordinary people to contribute to knowledge. In contrast, for "intuitive thinking" (*l'esprit de finesse*), "its principles are in common usage and before the eyes of everyone. One has neither to turn one's head nor to do violence to oneself; it is only a question of having good sight—but it must be good: for the principles are so diluted and in such a great number that it is almost impossible for it not to escape" (*"les principes [de l'esprit de finesse] sont dans l'usage commun et devant les yeux de tout le monde. On n'a que faire de tourner la tête, ni de se faire violence; il n'est question que d'avoir bonne vue, mais il faut l'avoir bonne: car les principes sont si déliés et en si grand nombre, qu'il est presque impossible qu'il n'en échappe"* [ibid., pp. 244–45]). Unlike the mathematical or geometric approach characteristic of science, the reasoning of *l'esprit de finesse* involves judgment (Pensée 513, p. 245). Individuals differ in the capacity that Pascal, metaphorically, calls "good" sight or vision. The methods brought to perfection by ancient science are difficult because only a few individuals can have them to the highest degree.

19. Even more difficult, according to Pascal, is the combination of these two modes of thought. "And thus it is rare that the mathematicians are intuitive and the intuitive are mathematicians, because the mathematicians want to treat mathematically the intuitive things and make themselves ridiculous, wanting to begin by definitions and then by principles, which is not the manner of acting in this kind of reasoning" (*"Et ainsi il est rare que les géomètres soit fins et que les fins soient géomètres, à cause que les géomètres veulent traiter géometriquement ces choses fines et se rendent ridicules, voulant commencer par les définitions et ensuite par les principes, ce qui n'est pas la manière d'agir en cette sorte de raisonnement"* [ibid., p. 245]). Limiting all reasoning to the approach of Bacon, Hobbes, and extension of

the modern scientific method to human things, necessarily impoverishes the subject matter of human life.

Those mathematicians who are only mathematicians thus have correct thinking, but only provided that one explains to them all things by definitions and principles; otherwise they are false and insupportable, for they are only correct when based on well clarified principles. And the intuitive thinkers who are only intuitive cannot have the patience to follow down to the first principles speculative and imaginary things that they have never seen in the world and which are entirely out of use" (*Les géomètres, qui ne sont que géomètres ont donc l'esprit droit, mais pourvu qu'on leur explique bien toutes choses par définitions et principes; autrement ils sont faux et insupportables, car ils ne sont droits que sur les principes bien éclaircis. Et les fins qui ne sont que fins ne peuvent avoir la patience de descendre jusques dans les premiers principes des choses spéculatives et d'imagination qu'ils n'ont jamais vues dans le monde, et tout à fait hors d'usage*). (Pensée 512, p. 246).

20. C. S. Lewis, *The Abolition of Man* (New York: Collier, 1962), p. 69; Leon Kass, "The New Biology: What Price Relieving Man's Estate," *Science* 174 (1971): 779–88.

21. Cf. ibid., esp. chap. 2, with James Q. Wilson, *On Character* (Washington, D.C.: AEI Press, 1991). The application of this perspective is beautifully exemplified by George Anastaplo, *Human Being and Citizen: Essays on Freedom, Virtue, and the Common Good* (Chicago: Swallow Press, 1975). For the view that the defense of virtue or "character" in a time of social change is at the heart of Plato's thought, see Irving M. Zeitlin, *Plato's Vision: The Classical Origins of Social and Political Thought* (New York: Prentice-Hall, 1993), esp. pp. 92 ff.

22. This definition can be put in "objective" terms by referring to the problem of selfishness and cooperation posed by the Prisoners' Dilemma in game theory (e.g., Robert Axelrod, *The Evolution of Cooperation* [Cambridge, Mass.: Harvard University Press, 1983]). When two rational, profit-maximizing individuals cannot communicate with each other, the "friend" will initiate and persist in cooperative behavior even in the absence of information indicating the partner's reciprocity. Mutual benefit can be achieved by what Axelrod calls the "TIT-for-TAT" strategy in a repeated Prisoner's Dilemma: start by cooperating but then behave exactly as the other partner did in the previous round. Such a strategy, characteristic of the *lex talionis* in societies without governments ("an eye for an eye"), is less altruistic than the virtue we seek in a friend.

23. For the traditional understanding of this issue, see Xenophon's *Hiero*, in *On Tyranny*, rev. ed., ed. Leo Strauss (Glencoe, Ill.: Free Press, 1966), pp. 1–20; Aristotle, *Nicomachean Ethics*, 8.1.1155a–9.12.1172a; and Montaigne's "Apology of Raymond Sebond," in *Complete Essays*, ed. Donald M. Frame (Stanford, Calif.: Stanford University Press, 1965).

24. Rousseau put it well in contrasting "politeness" (the mutually beneficial conventions of routine society, which do not have a net cost for either partner) from true friendship: "Incessantly politeness requires, propriety demands; incessantly usage is followed . . . . Therefore one will never know well those with whom he deals, for to know one's friend thoroughly, it would be necessary to wait for emergencies—that is, to wait until it is too late" (*Discourse on the Sciences and Arts*, in *Collected Writings of Rousseau*, ed. Roger D. Masters and Christopher Kelly [University Press of New England, 1992], vol. 2, p. 6).

25. For Aristotle and other ancients, it is necessary to distinguish practical or moral virtues from the intellectual virtues: while knowledge of nature is related to good action, it is easy to imagine scientists who intentionally use their expertise in the service of evil purposes. Indeed, one of Aristotle's most frequent criticisms of Socrates is the Socratic tendency to equate knowledge with the practical virtues as a whole; e.g., *Nicomachean Ethics*, 6.13.1144b: "Socrates in one respect was on the right track, while in another he went astray; in thinking that all the excellences were forms of practical wisdom he was wrong, but in saying they implied practical wisdom he was right." See also *Nicomachean Ethics*, 3.7.1116b, 7.2.1145b, 7.3.1147b; *Magna Moralia*, 1.1.1182a.

26. The example of Nazism and the "final solution" comes immediately to mind. Is it entirely an accident that some of those most influential in outlining the principles of modern relativism—notably Carl Schmidt, Martin Heidegger, and Paul DeMan—openly supported fascism under Hitler? History can hardly provide an assurance of "progress" toward more civilized norms. Cf. Wolfgang Fikentscher, "The Sense of Justice and the Concept of Cultural Justice: Legal Anthropology," in *The Sense of Justice*, ed. Roger D. Masters and Margaret Gruter, (Newbury Park, Calif.: Sage, 1992), pp. 106–27.

27. For the evidence and implications of this statement, see the essays by Michael T. McGuire, Robert Frank, Margaret Gruter, Herbert Helmrich, and Frans B. M. de Waal in Masters and Gruter, eds., *The Sense of Justice* chaps. 2–3, 5, 10–11.

28. On natural differences in ability or potential, see esp. Howard Gardner, *Frames of Mind* (New York: Basic Books, 1983). The self-imposed obligations to fulfill these abilities is beautifully expressed in the concept of the "noble" individual in José Ortega y Gasset, *The Revolt of the Masses* (New York: W. W. Norton, 1932).

29. Plato, *Republic*, 618b–e (ed. Allan Bloom [New York: Basic Books, 1968], p. 301).

Index

Page numbers in **boldface** indicate illustrations.

UNIVERSITY PRESS OF NEW ENGLAND publishes books under its own imprint and is the publisher for Brandeis University Press, Brown University Press, University of Connecticut, Dartmouth College, Middlebury College Press, University of New Hampshire, University of Rhode Island, Tufts University, University of Vermont, and Wesleyan University Press.

LIBRARY OF CONGRESS CATALOGING-IN-PUBLICATION DATA

Masters, Roger D.

Beyond relativism : science and human values / Roger
D. Masters.

p.   cm.

Includes bibliographical references and index.

ISBN 0–87451–634–X

1. Methodology.   2. Science—Philosophy.   3. Social sciences—
Philosophy.   4. Ethics.   5. Humanities.   6. Relativity.   I. Title.

B53.M36   1993

001—dc20   93–16925

♾